Progress in Mathematics
Volume 67

Discrete Groups in Geometry and Analysis

Papers in Honor of G.D. Mostow on His Sixtieth Birthday

Edited by
Roger Howe

1987

Birkhäuser
Boston · Basel · Stuttgart

Roger Howe
Department of Mathematics
Yale University
New Haven, CT 06520
U.S.A.

Library of Congress Cataloging in Publication Data
Discrete groups in geometry and analysis.
(Progress in mathematics ; v. 67)
Papers from a conference held Mar. 23–25, 1986, at
Yale University.
1. Discrete groups—Congresses. 2. Mostow, George D.
I. Howe, Roger. II. Mostow, George D. III. Series.
QA171.D565 1987 512′.22 86-29955

CIP-Kurztitelaufnahme der Deutschen Bibliothek
Discrete groups in geometry and analysis : papers
in honor of G.D. Mostow on his 60. birthday /
Roger Howe, ed.—Boston : Basel : Stuttgart :
Birkhäuser, 1987.
(Progress in mathematics : Vol. 67)
ISBN 3-7643-3301-4 (Basel . . .)
ISBN 0-8176-3301-4 (Boston)
NE: Howe, Roger [Hrsg.]; Mostow, George D..
Festschrift; GT

ISBN 0-8176-3301-4
ISBN 3-7643-3301-4

Printed and bound by R.R. Donnelley & Sons, Harrisonburg, Virginia.
Printed in the U.S.A.

9 8 7 6 5 4 3 2 1

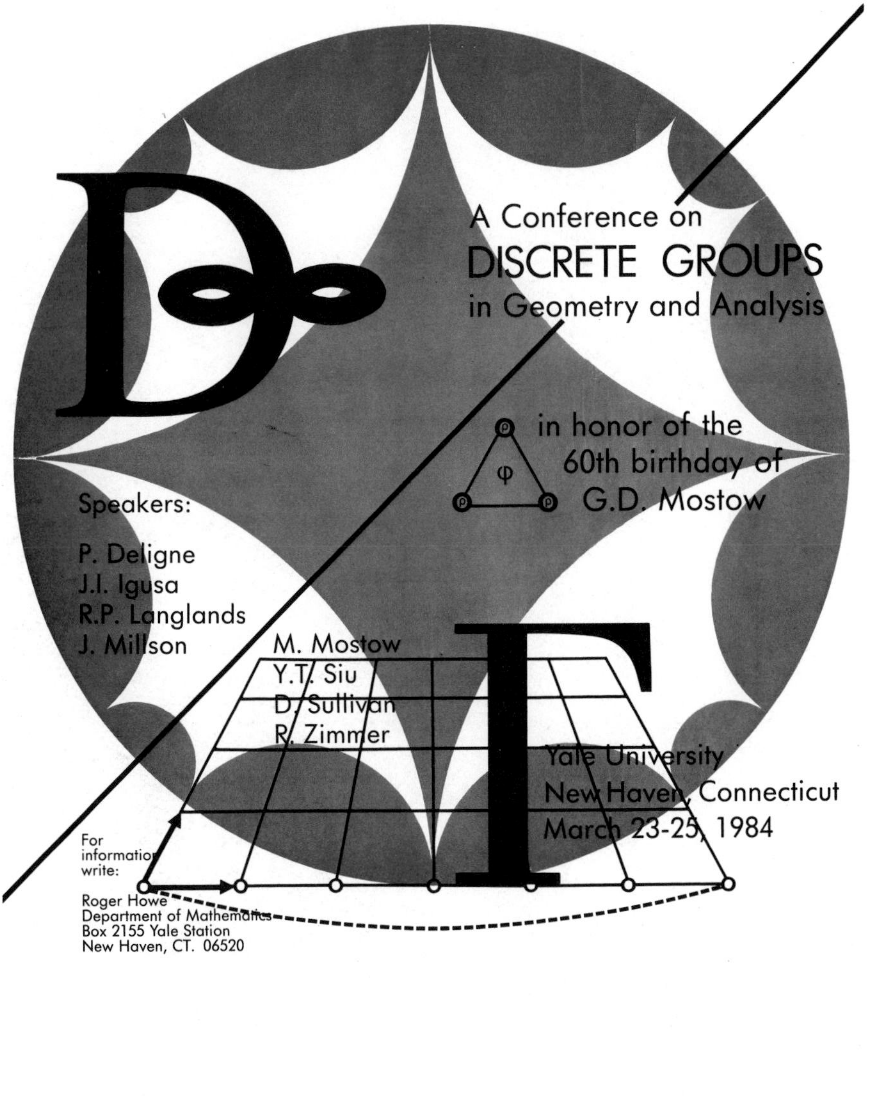
D
A Conference on
DISCRETE GROUPS
in Geometry and Analysis
in honor of the
60th birthday of
G.D. Mostow
Speakers:
P. Deligne
J.I. Igusa
R.P. Langlands
J. Millson
M. Mostow
Y.T. Siu
D. Sullivan
R. Zimmer
Yale University
New Haven, Connecticut
March 23-25, 1984
For information write:
Roger Howe
Department of Mathematics
Box 2155 Yale Station
New Haven, CT. 06520

PREFACE

On the weekend of March 23-25 we held at Yale a conference to mark the 60th birthday of Dan Mostow. Eight talks were given, by Pierre Deligne, Jun-Ichi Igusa, Robert Langlands, John Millson, Mark Mostow, Yum-Tong Siu, Dennis Sullivan, and Robert Zimmer. Of the many pieces of evidence one might cite to indicate Dan's stature in the mathematical community, the one I appreciated most as principal organizer of the conference was the ease with which this roster of speakers was assembled. Their well-crafted lectures assured the mathematical success of the conference.

Six of the eight speakers further agreed to submit manuscripts based on their talks. This volume is the result. As the reader will see, the articles here are not simply technical reports or research summaries, but contain well-organized developments of significant mathematics, much of it appearing here for the first time. They represent substantial work, as is appropriate for the man they commemorate.

The conference and this book were made possible through the cooperation of many people. I would like to take this chance to acknowledge some of them here. First, thanks to the speakers and authors, named above. I want also to express appreciation to the mathematicians who were kind enough to read and comment on the manuscripts for the book: William Goldman, Robert Greene, Diane Meuser, and Gopal Prasad. The NSF underwrote many of the expenses of organizing the conference and of producing the book. For cheerful help with registration, thanks to Bernadette Highsmith. Donna Belli and Mel DelVecchio rapidly and skillfully produced multiple typescripts for the papers appearing here. Alan Durfee added a nice touch to the conference by playing the Yale Carillon on Saturday. And finally, thanks to the people at Birkhauser, which in the first place meant Sigurdur Helgason, for their interest in and patience with this project.

Roger Howe
New Haven
October 1986

G.D. Mostow
(Photo courtesy of T. Charles Erickson, Yale University)

BIOGRAPHICAL SKETCH

George D. "Dan" Mostow is Henry Ford II Professor of Mathematics at Yale University. He has been a member of the National Academy of Sciences since 1974. He will be President of American Mathematical Society in 1987-88.

Dan was born July 4, 1923, attended Boston Latin School and Harvard College, graduating in 1943. His Ph.D. is also from Harvard (1948) and was done under the guidance of G.D. Birkhoff. He has taught at Princeton (1947-48), Syracuse University (1949-52), Johns Hopkins University (1952-61) and has been at Yale since 1961. He served as Chairman of the Mathematics Department from 1971 to 1974.

Dan has been a Member of the Institute for Advanced Study in Princeton three times, in 1947-49, 1956-57, and 1975. He has been a Visiting Professor at the Instituto de Matematico in Rio de Janeiro (1953-54), at the Institut des Hautes Etudes Scientifiques in Bures-sur Yvette, France (1966, 71, and 75) and the University of Paris (1966-67), at the Hebrew University in Jerusalem (1967), and at the Tata Institute of Fundamental Research, Bombay, India (1970). He was a John Simon Guggenheim Fellow and a Fulbright Scholar at the Mathematics Institute in Utrecht, Netherlands (1957-58).

Dan has served on about a dozen committees of the American Mathematical Society. He has been editor (1965-69) and associate editor (1969-) of the American Journal of Mathematics. He has also been Associate Editor of the Annals of Mathematics (1957-64), of the Transactions of the American Mathematical Society (1958-65) and of American Scientist (1970-82). With Armand Borel he organized the Summer Institute of the American Mathematical Society in Boulder, Colorado in 1965, and they edited the resulting volume of proceedings (Algebraic Groups and Discontinuous Subgroups, Proceedings of Symposia in Pure Mathematics, volume IX). He chaired the U.S. National Committee for Mathematics in 1971-73 and 1983-85, and the Office of Mathematical Sciences of the National Academy of Sciences - National Research Council in 1975-78. He has been a trustee of the Institute for Advanced Study since 1982. He helped form and served on the Ad Hoc Committee on Resources for the Mathematical Sciences (1981-84), which issued the report Renewing U.S. Mathematics.

Dan's research has mostly concerned the geometry of Lie groups, especially discrete subgroups of Lie groups. He has authored over 60 books and papers, including the monograph Strong Rigidity of Locally Symmetric Spaces, Annals of Mathematics Studies 78, Princeton University Press, 1973, which figures prominently in this volume.

Dan lives with his wife Evelyn in a house of his own design. They have four grown children: Mark, Jack, Carol, and Jonathan; and three grandchildren.

TABLE OF CONTENTS

UN THÉORÈME DE FINITUDE POUR LA MONODROMIE

par P. Deligne

0. Introduction.

Soit S une variété algébrique complexe lisse (i.e. sans singularité) connexe et soit $(X_s)_{s\in S}$ une famille algébrique, paramétrisée par S, de variétés projectives lisses $X_s \subset \mathbb{P}^N(\mathbb{C})$. Par définition, c'est la donnée d'un morphisme projectif et lisse $f : X \to S$, muni d'une factorisation

$$\begin{array}{ccc} X & \hookrightarrow & \mathbb{P}^N \times S \\ & {\scriptstyle f}\searrow \quad \swarrow{\scriptstyle pr_2} & \\ & S & \end{array}$$

et les X_s sont les fibres de $f : X_s := f^{-1}(s)$. Quel que soit i, les groupes de cohomologie $H^i(X_s,\mathbb{Z})$ forment un système local sur S. Si on choisit un point base $o \in S$, le groupe fondamental $\pi_1(S,o)$ agit donc sur $H^i(X_o,\mathbb{Z})$. C'est la représentation de monodromie

$$\sigma_{\mathbb{Z}} : \pi_1(S,o) \to \mathrm{Aut}(H^i(X_o,\mathbb{Z})).$$

L'indice $\mathbb{Z}$ rappelle que $\pi_1(S,o)$ agit sur un $\mathbb{Z}$-module de type fini. Nous considérerons surtout la représentation correspondante sur le $\mathbb{Q}$-espace vectoriel $H^i(X_o,\mathbb{Q}) = H^i(X_o,\mathbb{Z}) \otimes \mathbb{Q}$, encore dite de monodromie, et notée simplement

$$\sigma : \pi_1(S,o) \to GL(H^i(X_o,\mathbb{Q})).$$

On ignore à peu près tout de quels groupes peuvent être groupes fondamentaux de variétés algébriques, et de quelles représentations linéaires peuvent être obtenue comme monodromie. Par exemple, on ne connaît pas la résponse à la question de Serre: $\pi_1(S,o)$ peut-il être non-trivial, et n'avoir aucon quotient fini non-trivial? Une condition nécessaire pour qu'un groupe soit groupe fondamental de variété algébrique a été donnée par J. Morgan [8].

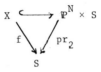

L'adhérence de Zariski M^- du <u>groupe de monodromie</u> $M := \mathrm{Im}(\sigma)$ dans le groupe linéaire $GL := GL(H^i(X_o,\mathbb{Q}))$ est souvent plus facile à calculer que $\mathrm{Im}(\sigma)$. On sait que c'est un groupe algébrique semi-simple ([2], 4.2.9a)). Par abus de notation, notons $M^-(\mathbb{Z})$ le sous-groupe de $M^-(\mathbb{Q})$ qui respecte le réseau entier $H^i(X_o,\mathbb{Z})/\text{torsion} \subset H^i(X_o,\mathbb{Q})$. On a $M \subset M^-(\mathbb{Z})$. On connaît des exemples où M est d'indice infini dans $M^-(\mathbb{Z})$ (Mostow-Deligne [3], M. Nori [9]). Dans l'exemple de Nori, le groupe M n'est pas de présentation finie. Que, initialement, les exemples qu'on avait pu calculer fournissaient tous des groupes M d'indice fini dans $M^-(\mathbb{Z})$ provient peut-être de ce qu'on manque de critère pour reconnaître le contraire. On ignore si, pour $\gamma \in M^-(\mathbb{Z})$, la question de savoir si $\gamma \in M$ est décidable.

Cet article expose quelques résultats de finitude sur les représentations de monodromie. Le plus frappant est le

<u>Théorème 0.1</u>. <u>Fixons</u> S, $o \in S$ <u>et un entier</u> N. <u>Les représentations de monodromie</u> σ <u>de</u> $\pi_1(S,o)$ <u>(pour</u> i <u>et</u> X <u>variables) qui sont de dimension</u> N <u>ne forment qu'un nombre fini de classes d' isomorphie</u>.

Nous prouverons ausi le résultat plus précis:

<u>Variante 0.2</u>. <u>Fixons</u> S, $o \in S$ <u>et un entier</u> N. <u>Les représentations rationnelles de dimension</u> N <u>de</u> $\pi_1(S,o)$, <u>qui sont facteurs directs d'une représentation de monodromie ne forment qu'un nombre fini de classes d'isomorphie</u>.

Ces énoncés sont pour les représentations <u>rationnelles</u> fournie par la monodromie. Les lemmes bien connus suivant permettent de traiter aussi de représentations sur les entiers.

<u>Lemme 0.3</u>. <u>Soit</u> $\sigma:\Gamma \to GL(V)$ <u>une représentation semi-simple d'un groupe</u> Γ <u>sur un espace vectoriel rationnel de dimension finie</u> V. <u>Les représentations</u> $\Gamma \to \mathrm{Aut}(L)$, <u>pour</u> $L \subset V$ <u>un réseau entier stable par</u> Γ, <u>ne forment qu'un nombre fini de classes d'isomorphie</u>.

La preuve est rappelée en 2.6.

<u>Lemma 0.4</u>. <u>Soient</u> H <u>un</u> $\mathbb{Z}$<u>-module de type fini</u>, T <u>son sous-groupe de torsion</u>, $L = H/T$ <u>et</u> $\sigma:\Gamma \to \mathrm{Aut}(L)$ <u>une représentation d'un groupe</u> Γ <u>sur</u> L. <u>Si</u> Γ <u>est de type fini, il n'y a qu'un nombre fini de relèvements de</u> σ <u>en une action de</u> Γ <u>sur</u> H.

<u>Preuve</u>. Le noyau de $\mathrm{Aut}\, H \to \mathrm{Aut}\, L$ est fini, car contenu dans

l'ensemble fini $1 + \mathrm{Hom}(H,T)$. Pour chaque générateur γ de Γ, $\sigma(\gamma)$ n'a donc qu'un nombre fini de relèvements dans $\mathrm{Aut}\, H$.

Si $f : X \to S$ est un morphisme projectif et lisse, on dispose pour chaque X_s d'une décomposition de Hodge

$$H^i(X_s,\mathbb{Q}) \otimes \mathbb{C} = H^i(X_s,\mathbb{C}) = \bigoplus_{p+q=i} H^{p,q}.$$

Le sous-espace $H^{p,q}$ de $H^i(X_s,\mathbb{C})$ est l'espace des classes de cohomologie c représentables par une forme fermée de type (p,q). Pour toute structure kahlérienne sur X, il revient au même d'exiger que le représentant harmonique de c soit de type (p,q). Quand s varie, les $H^i(X_s,\mathbb{C})$ forment un système local sur S, et la décomposition de Hodge varie de facon analytique réelle. P.A. Griffiths a dégagé des propriétés essentielles de comment la décomposition de Hodge varie avec s. Si on les prend comme axiomes, on obtient la notion de "variation polarisable de structures de Hodge de poids i": un système local de $\mathbb{Z}$-modules de type fini, de complexifiés munis d'une décomposition "de Hodge" variant continûment et satisfaisant à des axiomes convenables ([6], 12). Nous considérerons surtout le système local de $\mathbb{Q}$-espaces vectoriels de dimension finie déduit du système local de $\mathbb{Z}$-modules de type fini en tensorisant avec $\mathbb{Q}$. Nous dirons qu'il est sous-jacent à la variation.

Le théorème 0.1 et sa variante 0.2 résultent du

Théorème 0.5. Fixons S et un entier N. Les systèmes locaux sur S de $\mathbb{Q}$-espaces vectoriels de dimension N, facteurs directs d'un système local sous-jacent à une variation de structures de Hodge polarisable, ne forment qu'un nombre fini de classes d'isomorphie.

La preuve est donnée au paragraphe 2. Le lecteur intéressé seulement par 0.1, ou ne tenant pas dans 0.5 à pouvoir prendre des facteurs directs, peut omettre la fin du paragraphe 1, à partir de 1.11.

Le présent travail est le fruit d'un effort pour comprendre l'article [4] de G. Faltings. Au paragraphe 3 on explique brièvement comment, dans le cas où S est une courbe, où X est un schéma abélien sur S et où $i = 1$, le théorème 0.1 peut être obtenu par les méthodes de [4]. On obtient aussi un théorème "uniforme en S" pour S une courbe de type topologique donné (3.10). J'ignore si ce théorème vaut pour des variations de structures de Hodge polarisables de type plus général que $\{(0,1),(1,0)\}$.

1. Variations Complexes.

1.1. Soit S une variété analytique complexe lisse. Une variation complexe de poids 0 sur S est la donnée d'un système local complexe V sur S, de fibres V_s $(s \in S)$ munies d'une décomposition $V_s = \bigoplus_{p \in \mathbb{Z}} V_s^p$, satisfaisant aux axiomes suivants.

(V.C.1) Les sous-espaces $F^p := \bigoplus_{i \geq p} V_s^i$ (resp. les $\bar{F}^q := \bigoplus_{i \leq -q} V_s^i$) de V_s varient holomorphiquement (resp. antiholomorphiquement) avec s.

(V.C.2) Si une section locale différentiable v de V est en chaque point dans F^p (resp. $\bar{F}^q$), sa dérivée par rapport à un champ de vecteurs sur S est dans F^{p-1} (resp. $\bar{F}^{q-1}$).

Si $\mathcal{V}$ est le faisceau des sections C^∞ de V et que ∇ est la connexion, vue comme morphisme de faisceaux $\nabla : \mathcal{V} \to \Omega^1(\mathcal{V})$, les décompositions des V_s en les V_s^p fournissent une décomposition $\mathcal{V} = \bigoplus \mathcal{V}^p$ de $\mathcal{V}$, et (V.C.1), (V,C,2) équivalent à ce que ∇ envoie $\mathcal{V}^i$ dans $\Omega^{1,0}(\mathcal{V}^{i-1}) \oplus \Omega^1(\mathcal{V}^i) \oplus \Omega^{0,1}(\mathcal{V}^{i+1})$.

Dans la suite, nous ne considérerons que des variations complexes de poid 0 et omettrons de préciser "de poids 0".

Pour S connexe, les nombres de Hodge h^p sont les dimensions $h^p := \dim V_s^p$, indépendantes de $s \in S$. De même, la dimension de V est celle des V_s.

Une polarisation de la variation complexe V est une forme hermitienne horizontale ψ sur V telle qu'en chaque point $s \in S$ la décomposition $V = \bigoplus V_s^p$ soit orthogonale, et que la restriction de ψ à V_s^p soit définie positive pour p pair et définie négative pour p impair. On posera $\Phi = \Sigma(-1)^p \psi | V^p$, C'est une forme hermitienne définie positive sur le fibré vectoriel $\mathcal{V}$.

1.2. Exemple: Soient H une variation de structures de Hodge de poids w sur S et V le complexifié du système local d'espaces vectoriels rationnels $H_{\mathbb{Q}}$ sous-jacent à H. Par définition, il est muni de décompositions

$$V_s = \bigoplus_{p+q=w} V_s^{pq} \qquad (s \in S).$$

Si on pose $V_s^p := V_s^{p,w-p}$, V, muni de ces décompositions, est une variation complexe. Une polarisation ψ_o de H est une forme bilinéaire horizontale sur $H_{\mathbb{Q}}$ à valeurs dans $(2\pi i)^{-w}\mathbb{Q}$ vérifiant des axiomes convenables. Si ψ est le prolongement sesquilinéaire de ψ_o à V, ces axiomes équivalent à ce que $(2\pi i)^w \psi$ soit une polarisation de la variation complexe V.

1.3. Spécialisons 1.1 au cas où S est réduit à un point: une <u>structure de Hodge complexe</u> sur un espace vectoriel complexe H est une décomposition $H = \bigoplus_{p \in \mathbb{Z}} H^p$; ses <u>nombres de Hodge</u> sont les $h^p := \dim H^p$; une <u>polarisation</u> est une forme hermitienne ψ pour laquelle les H^p sont deux à deux orthogonaux et de restriction à H^p définie positive pour p pair, définie négative pour p impair.

Soient H un espace vectoriel complexe, ψ une forme hermitienne sur H et (h^p) une famille d'entiers ≥ 0. On suppose que $\dim H = \Sigma h^p$ et que $\mathrm{sgn}(\psi) = \Sigma(-1)^p h^p$. On notera $M(H,\psi,(h^p))$ ou simplement M l'espace des structures de Hodge complexes sur H, de nombres de Hodge h^p, polarisées par ψ. L' application qui à un point de M associe la <u>filtration de Hodge</u> de H par les $F^p := \bigoplus_{i \geq p} H^i$ est une injection de M dans une variété de drapeaux de H. En effet, $\bar{F}^q := \bigoplus_{i \leq -q} H^i$ est l'orthogonal de F^{1-q} et $H^p = F^p \cap \bar{F}^{-p}$. Cette injection identifie M à un ouvert de cette variété de drapeaux. On munit M de la structure complexe induite.

En chaque point de M, End(H) hérite d'une décomposition de Hodge, et d'une filtration de Hodge correspondante, avec

$$F^p\mathrm{End}\, H = \{f: H \to H \mid f(F^i) \subset F^{i+p}\}.$$

L'espace tangent à M (i.e. à la variété de drapeaux) s'identifie à $\mathrm{End}(H)/F^o\mathrm{End}(H)$, et hérite de la filtration image de F. C'est une filtration holomorphe F du fibré tangent, avec $F^o = 0$.

Soit U le groupe unitaire $U(H,\psi)$. Il agit transitivement sur M. Si $m \in M$ correspond à une décomposition $H = \oplus H^p$ de H, le stabilisateur de m est le sous-groupe compact

$$K := \Pi\, U(H^p, \psi|H^p) \subset U.$$

La compacité de K assure l'existence sur $M \approx U/K$ de métriques riemanniennes (voire hermitiennes) U-invariantes.

<u>1.4. Remarque</u>: Si (H',ψ') est isomorphe à (H,ψ), une métrique U-invariante sur M en détermine une sur $M' := M(H',\psi',(h^p))$: transporter la métrique donnée par l'isomorphisme $M \to M'$ induit par un isomorphisme $u:(H,\psi) \to (H',\psi')$. La U-invariance assure que la métrique obtenue sur M' ne dépend pas du choix de u.

<u>1.5. Remarque</u>: Pour toute métrique riemannienne U-invariante d sur

M, la variété riemannienne (M,d) est homogène; elle est donc complète et ses boules fermées sont compactes. Si $o \in M$ et que C est une constante, l'ensemble des $g \in U$ tels que $d(go,o) \leq C$ est compact.

1.6 Soient (V,ψ) une variation complexe polarisée sur S, supposé connexe, et (h^p) ses nombres de Hodge. Choisissons un point base $o \in S$ et soit $(\tilde{S},o)$ le revêtement universel de (S,o). Notons V_o la fibre du système local V en o et ψ_o la forme hermitienne sur V_o induite par ψ. L'image inverse sur $\tilde{S}$ de la variation complexe V est une variation complexe sur $\tilde{S}$. Le système local sous-jacent est constant. Il s'identifie au système local constant de fibres V_o et cette identification permet de regarder la variation comme une famille, paramétrisée par $\tilde{S}$, de structures de Hodge complexes sur l'espace vectoriel fixe V_o. Elles sont polarisées par ψ_o et de nombres de Hodge (h^p). On attache ainsi à (V,ψ) une application

$$P : \tilde{S} \to M(V_o,\psi_o,(h^p)).$$

L'axiome (V.C.1) signifie que P est une application holomorphe, et (V.C.2) que dP prend ses valeurs dans le sous-fibré F^{-1}(tangent) du fibré tangent (cf. 1.3).

On dispose d'une action de $\pi_1(S,o)$ sur $\tilde{S}$ et sur V_o (monodromie) - donc sur $M := M(V_o,\psi_o,(h^p))$ - et l'application P est équivariante. L'application P détermine la variation complexe image inverse de (V,ψ) sur $\tilde{S}$, et la variation complexe (V,ψ) sur S s'en déduit par passage au quotient par $\pi_1(S,o)$.

Notre outil essentiel est le théorème suivant de P.A. Griffiths ([5], 10.1).

1.7. Théorème. (P.A. Griffiths). Supposons que S est le disque unité, et munissons-le de sa métrique de Poincaré (courbure constante -1). Alors, pour une métrique riemannienne U-invariante convenable sur M, ne dépendant que des nombres de Hodge h^p (cf. 1.4), l'application P décroît les distances.

Reprenons les hypothèses et notations de 1.6, et soit $\sigma:\pi_1(S,o) \to GL(V_o)$ la représentation de monodromie. Nous dirons qu'une base e de V_o est adaptée à la décomposition de Hodge $P(o) : V_o = \oplus V_o^p$ de V_o si elle est la réunion de bases orthonormées pour $(-1)^p\psi$ des V_o^p.

1.8. Corollaire. Fixons (S,o) et les nombres de Hodge (h^p). Pour chaque $\gamma \in \pi_1(S,o)$, il existe C tel que, pour toute variation polarisée V sur S, de nombres de Hodge (h^p), et pour toute base adaptée e de V_o, les coefficients de la matrice $\sigma(\gamma)$, dans la base e, sont bornés en valeur absolue par C.

Preuve. Soient $H := \oplus\, \mathbb{C}^{h^p}$, ψ la forme hermitienne sur H somme des formes hermitiennes $(-1)^p \Sigma u_i \bar{v}_i$ sur les $\mathbb{C}^{h^p}$, $M := M(H,\psi,(h^p))$ et o le point de M correspondant à la décomposition de H en les $\mathbb{C}^{h^p}$. Pour toute variation V comme en 1.7, le choix d'une base adaptée de V_o identifie V_o à H et permet de regarder l'application P définie par V comme une application

$$P_V : \tilde{S} \to M$$

vérifiant $P_V(o) = o$.

Soit d_K la pseudo-métrique de Kobayashi sur $\tilde{S}$: la plus grande pseudo-métrique d telle que pour toute application holomorphe φ du disque unité dans $\tilde{S}$, $d(\varphi(x),\varphi(y))$ soit inférieur ou égal à la distance de Poincaré de x à y. Puisque $\tilde{S}$ est connexe, $d_K(x,y)$ est fini pour tout x et y. Pour M muni d'une métrique riemannienne U-invariante d_M convenable, indépendante de (S,o) et de V, il résulte de 1.7 que

$$d_M(\sigma(\gamma).o,o) = d_M(P_V(\gamma.o),P_V(o)) \le d_K(\gamma.o,o).$$

D'après 1.5, $\sigma(\gamma)$ est donc dans un compact de U indépendant de V, ce qui prouve 1.8.

1.9. Corollaire. Fixons (S,o) et un entier N. Pour chaque $\gamma \in \pi_1(S,o)$, il existe C tel que, pour toute variation polarisée V sur S, de dimension N, et toute base adaptée e de V_o, les coefficients de la matrice $\sigma(\gamma)$, dans la base e, sont bornés en valeur absolue par C.

Preuve. Procédons par récurrence sur N. Le cas $N = 0$ est trivial. Si p est tel que $h^p = 0$, soit $V_{<p}$ (resp. $V_{>p}$) la somme des V^i pour $i < p$ (resp. $i > p$). Il résulte de (V.C.1)(V.C.2) que les sous-fibrés $V_{<p}$ et $V_{>p}$ sont horizontaux. S'il existe $i < p < j$ avec $h^i \neq 0$, $h^p = 0$, $h^j \neq 0$, on obtient ainsi une décomposition de la variation complexe V en la somme de deux variations complexes de dimensions strictement plus petites que celle de V. Une base adaptée de V_o est réunion de bases adaptées de $(V_{<p})_o$ et $(V_{>p})_o$, et on conclut par récurrence.

Il reste à traiter le cas où les i tels que $h^i \neq o$ forment un intervalle. Soit $V(n)$ la variation complexe de même système local sous-jacent que V, déduite de V par la renumérotation: $(V(n)_s)^p := V_s^{p+n}$. Elle est polarisée par $(-1)^n\psi$, une base adaptée de V_o est encore une base adaptée de $V(n)_o$, et les coefficients de la matrice $\sigma(\gamma)$ sont les mêmes pour V et $V(n)$. On ne restréint donc par la généralité en ne traitant que des V tels que $\{p|h^p \neq 0\}$ soit un intervalle commencant en 0. Il n'y a alors qu'un nombre fini de possibilités pour le système des h^p, et on conclut par 1.8.

<u>1.10 Corollaire</u>. <u>Fixons</u> (S,o) <u>et un entier</u> N. <u>Pour chaque</u> $\gamma \in \pi_1(S,o)$, <u>il existe</u> C <u>tel que, pour toute variation complexe polarisable</u> V <u>sur</u> S, <u>de dimension</u> N, <u>on ait</u>

$$|\mathrm{Tr}(\sigma(\gamma))| < C .$$

1.11. Dans la fin de ce paragraphe, on suppose que S est le complément, dans une variété analytiqne compacte $\bar{S}$, d'un sous-espace analytique. Cette hypothèse implique que sur S toute fonction plurisubharmonique bornée supérieurement est constante. Soit V une variation complexe polarisable de structures de Hodge sur S. Si U est un ouvert de S, et v une section horizontale de V sur U, W. Schmid [10] montre que $\phi(v,v)$ est borné sur la trace sur U de tout compact de $\bar{S}$. Mise en garde: le cadre dans lequel travaille W. Schmid est différent du nôtre, mais ses preuves s'adaptent sans difficulté: il travaille avec des "variations réelles"; chaque variation complexe définit une variation réelle de dimension double, et on applique [10] à cette dernière; il suppose les moncdromies locales quasi-unipotentes, mais ne se sert pas réellement de cette hypothèse. Ceci acquis, les arguments de [6], 7.1 (cf. [10], §7) montrent que si v est une section globale de V sur S, les composantes de v dans les V^p sont encore horizontales.

1.12. Soit V_o un système local complexe sur S. On suppose V_o semi-simple: il admet une décomposition

(1.12.1) $$V_o = \bigoplus_{i\in I} S_i \otimes W_i$$

où les S_i sont des systèmes locaux irréductibles deux à deux non isomorphes et où les W_i sont des espaces vectoriels complexes non nuls. L'hypothèse que V_o est semi-simple est automatiquement remplie si V_o est sous-jacent à une variation polarisable de structures de Hodge (par. 1.11, les arguments de [2], 4.2.6 s'appliquent). M. Nori (non publié) a montré qu'elle l'est aussi si V_o est sous-jacent à une variation complexe polarisable et que $\bar{S}$ est kahlérienne ou algébrique (ou plus généralement si chaque classe $a \in H^1(\bar{S},\mathbb{R})$ est représentable par une 1-forme $\alpha + \bar{\alpha}$ avec α holomorphe).

1.13. Proposition. *Sous les hypothèses de 1.11 et 1.12, si V_o est sous-jacent à une variation complexe polarisable,*

(i) *Chaque S_i est sous-jacent a une variation complexe polarisable, unique à une renumérotation $p \mapsto p + n$ près.*

(ii) *Choisissons sur chaque S_i une variation complexe polarisable. Via (1.12.1), des structures de Hodge complexes sur les W_i fournissent alors une variation complexe polarisable sur V_o. Toute variation complexe polarisable sur V_o est ainsi obtenue.*

Preuve. La décomposition (1.12.1) fournit un isomorphisme

(1.13.1) $\mathrm{End}(V_o) = \Pi\ \mathrm{End}(W_i)$.

L'espace vectoriel $\mathrm{End}(V_o)$ est l'espace des sections globales horizontales du système local $\underline{\mathrm{End}}(V_o)$. Pour V_o sous-jacent à une variation polarisable V, $\mathrm{End}(V_o)$ hérite de la décomposition de Hodge des $\underline{\mathrm{End}}(V_s)$ $(s \in S)$ (1.11):

$$\mathrm{End}(V_o) = \oplus\ \mathrm{End}(V_o)^p.$$

Lemma 1.14. *Toute graduation de $\Pi\ \mathrm{End}(W_i)$, compatible à la structure d'algèbre, provient de graduations des W_i.*

Preuve. Regardons une graduation comme une action du groupe multiplicatif $\mathbb{G}_m$, $\lambda \in \mathbb{G}_m(\mathbb{C}) = \mathbb{C}^*$ agissant par multiplication par λ^n sur la composante de degré n (SGA3 I4.7.3). La composante neutre du groupe des automorphismes de l'algèbre $\Pi\ \mathrm{End}(W_i)$ est le quotient du groupe $\Pi GL(W_i)$ par son centre $\mathbb{G}_m^I$. Tonte extension centrale de $\mathbb{G}_m$ par $\mathbb{G}_m^I$ étant triviale (SGA3IX8.2), un morphisme $\mathbb{G}_m \to \mathrm{Aut}(\Pi\mathrm{End}(W_i))$ se relève en un morphisme de $\mathbb{G}_m$ dans $\Pi GL(W_i)$, i.e. en une graduation des W_i.

Preuve de 1.13 (suite). Pour V_o sous-jacent à une variation polarisable V, choisissons (1.14) des graduations des W_i telles que l'isomorphisme (1.13.1) soit compatible aux graduations. Si la droite $L_i \subset W_i$ est homogène, c'est l'image d'un projecteur $e \in \mathrm{End}(W)$ homogène de degré 0 et $S_i \otimes L_i \subset V_o$, isomorphe à S_i, est l'image d'un projecteur $e \subset \mathrm{End}(V) = \mathrm{End}(V_o)^o$. C'est donc une sous-variation complexe de V, un facteur direct en fait, et on en déduit l'existence d'une variation complexe polarisable à laquelle S_i soit sous-jacent.

Choisissons sur chaque S_i une variation complexe polarisable. Pour toute variation polarisable V sur V_o,

$$W_i = \mathrm{Hom}(S_i, V_o)$$

hérite de la variation polarisable $\underline{\mathrm{Hom}}(S_i, V)$ d'une décomposition de Hodge (1.11 appliqué à $\underline{\mathrm{Hom}}(S_i, V)$). L'isomorphisme

$$\oplus S_i \otimes \mathrm{Hom}(S_i, V_o) \to V_o$$

respecte les structures de Hodge, et (ii) en résulte. L'assertion d'unicité dans (i) est (ii) pour $V_o = S_i$.

1.15 Corollaire. *Soient S un ouvert de Zariski de $\bar{S}$ compact, $o \in S$ et un entier N. Pour tout $\gamma \in \pi_1(S, o)$ il existe C tel que, pour toute variation de structures de Hodge polarisable W sur S et tout facteur direct V du système local rationnel sous-jacent, si $\dim(V) = N$, la monodromie $\sigma(\gamma)$ vérifie $|\mathrm{Tr}(\sigma(\gamma))| < C$.*

Résulte de 1.10 et 1.13.

2. Preuve du théorème 0.5.

Nous nous appuierons sur le théorème classique suivant.

2.1. Théorème. *Soient Γ un groupe de génération finie et N un entier. Il existe une partie finie F de Γ telle que si deux représentations linéaires de dimension N de Γ sur un corps k de caractéristique o, de caractères χ_1 et χ_2, vérifient $\chi_1(\gamma) = \chi_2(\gamma)$ pour $\gamma \in F$, alors $\chi_1 = \chi_2$.*

Preuve. Un groupe de génération finie est de génération finie en tant que monoïde. Il suffit donc de prouver 2.1 dans le cas plus général où Γ est un monoide de génération finie et où les représentations sont à valeurs dans les matrices $N \times N$, non nécessairement inversible. Soit T une partie finie de Γ qui engendre Γ. On ne restreint pas la généralité en supposant que Γ est le monoïde libre engendré par T. Dans ce cas, la donnée d'une représentation de Γ équivaut à celle d'une famille de matrices $N \times N$ indexée par T: à ρ, attacher la famille des $\rho(t)$ $(t \in T)$.

Soient $X^t_{i,j}$ $(i,j \in [1,N], t \in T)$ des indéterminées et A l'algèbre de polynômes à $N^2|T|$ variables $\mathbb{Q}[X^t_{i,j}]$. C'est l'algèbre des fonctions polynômes sur la variété algébrique (un espace affine) qui paramètre les familles indexées par T de matrices $N \times N$. Soit τ la représentation $\Gamma \to M_N(A)$ pour laquelle $\tau(t) = (X^t_{i,j})_{i,j \in [1,N]}$. L'action par $X \to g X g^{-1}$ du groupe linéaire GL_N sur les matrices $N \times N$ fournit une action sur la $\mathbb{Q}$-algèbre A du groupe algébrique GL_N (sur $\mathbb{Q}$). Soit A^{GL_N} l'algèbre des invariants. Pour tout espace vectoriel V de dimension N sur k, la k-algèbre $A^{GL_N} \otimes_{\mathbb{Q}} k$ s'identifie à l'algèbre des fonctions polynômes GL(V)-invariantes de $|T|$ endomorphismes de V. Les éléments $Tr(\tau(\gamma))$ $(\gamma \in \Gamma)$ de A sont invariants.

2.2 Lemme (C. Procesi [13]). L'algèbre des invariants de GL_N dans A est engendrée par les $Tr(\tau(\gamma))$ $(\gamma \in \Gamma)$.

La preuve consiste à se ramener par polarisation à l'étude des invariants multilinéaires de n endomorphismes, à les interpréter comme invariants multilinéaires de n vecteurs et n covecteurs et à utiliser la description que H. Weyl donne de ceux-ci.

Preuve de 2.1 (fin). D'après Hilbert, l'algèbre des invariants de GL_N dans A est de type fini. Il existe donc une partie finie F de Γ telle que tout invariant soit un polynôme en les $Tr(\tau(f))$ pour $f \in F$. En particulier, pour tout $\gamma \in \Gamma$, il existe un polynôme à coefficients rationnels P_γ en des indéterminées x_f $(f \in F)$, tel que

$$Tr\ \tau(\gamma) = P_\gamma((Tr\ \tau(f))_{f \in F}).$$

Pour tout corps k de caractéristique 0 et toute représentation de Γ dans $M_N(k)$, de caractère χ, on a par spécialisation,

$$\chi(\gamma) = P_\gamma((\chi(f))_{f \in F}).$$

Le théorème en résulte.

2.3 Remarque. Soit $s = s(N)$ le plus petit entier t tel que toute Q-algèbre associative sans unité vérifiant l'identité $z^N = 0$ vérifie aussi l'identité $z_1 \cdots z_t = 0$. Dans [13], C. Procesi montre que l'algèbre A^{GL_N} de 2.2 est engendrée par les $Tr(\tau(\gamma))$ pour γ de longueur $\leq s$, et que cette borne est optimale: si $|T| \geq s$ et que $\pi \in \Gamma$ est un produit de s générateurs distincts, $\mathrm{Tr}(\tau(\pi))$ n'est pas dans l'algèbre engendrée par les $\mathrm{Tr}(\tau(\gamma))$ pour γ de longueur $< s$. Dans 2.1, on peut donc prendre pour F l'ensemble des mots de longueur $\leq s$ en les éléments d'un système générateur symétrique T. La restriction "symétrique" est en fait inutile cas un caractère sur un groupe Γ est déterminé par sa restriction à un sous-monoïde qui engendre Γ.

Highman a montré que $s \leq 2^N - 1$. Pour une preuve très courte, vois N. Jacobson, Structure of rings (2nd ed.), p. 274. Cette borne a été améliorée par Yu. P. Razmislov en $s \leq N^2$ (Izvestia A.N. <u>38</u> 4 (1974), p. 756).

<u>2.4 Preuve de 0.5</u>. Fixons $o \in S$. Si V est une variation polarisable de structures de Hodge sur S, et que W est un facteur direct de dimension N de la représentation de monodromie σ correspondante, la représentation rationnelle σ_W de $\pi_1(S,o)$ sur W vérifie (A)(B) (C) ci-dessous.

(A) Pour tout $\gamma \in \pi_1(S,o)$, $\mathrm{Tr}(\sigma_W(\gamma)) \in \mathbb{Z}$.

<u>Preuve</u>. La représentation σ respecte par hypothèse un réseau entier. La sous-représentation σ_W aussi.

(B) Pour tout $\gamma \in \pi_1(S,o)$ il existe $C(\gamma,N)$ tel que

$$|\mathrm{Tr}(\sigma_W(\gamma))| < C(\gamma,N).$$

C'est une application de 1.15.

(C) La représentation σ_W est semi-simple.

En effet, σ l'est (cf. [2] 4.2.6, dont la méthode s'applique par 1.11).

A N fixé, par (A)(B), il n'y a qu'un nombre fini de possibilités pour la valeur de chaque $\mathrm{Tr}(\sigma_W(\gamma))$. D'après 2.1, il n'y a qu'un nombre fini de possibilités pour le caractère de σ_W, donc, d'après (C), pour sa classe d'isomorphie.

<u>2.5 Remarque</u>. Dans 0.5, on suppose que S est une variété algébrique. La preuve s'applique encore si S est un ouvert de Zariski d'une variété analytique compacte. Si on suppose seulement que S est une variété analytique dont le groupe fondamental est de génération finie, les mêmes arguments donnent encore que les variations de structures de Hodge polarisables de dimension N sur S ne donnent lieu qu'à un nombre fini de caractères de $\pi_1(S,o)$.

2.6. Preuve du Lemme 0.3. Il s'agit de prouver l'énoncé suivant.

Lemme: Soient $H = \mathbb{Z}^N$, Γ un groupe, σ une représentation $\Gamma \to \mathrm{Aut}(H)$ et supposons que l'action de Γ sur $H_{\mathbb{Q}} := H \otimes_{\mathbb{Z}} \mathbb{Q}$ soit complètement réductible. Soit G le groupe des automorphismes de $H_{\mathbb{Q}}$ qui commutent à l'action de Γ. Alors, les réseaux $H' \subset H_{\mathbb{Q}}$ stables par Γ ne forment qu'un nombre fini de G-orbites.

Preuve. Soient $A \subset \mathrm{End}(H)$ la sous-algèbre engendrée par les $\sigma(\gamma)$ $(\gamma \in \Gamma)$, et $A_{\mathbb{Q}} = A \otimes_{\mathbb{Z}} \mathbb{Q}$. La complète réductibilité de $H_{\mathbb{Q}}$ équivaut à ce que $A_{\mathbb{Q}}$ soit une algèbre semi-simple. Soit $A_1 \subset A_{\mathbb{Q}}$ un ordre maximal contenant A. Pour tout nombre premier ℓ, une étude locale montre que $G(\mathbb{Q}_\ell)$ agit transitivement sur les $\mathbb{Z}_\ell$-réseaux de $H \otimes \mathbb{Q}_\ell$ stables par $A_1 \otimes \mathbb{Z}_\ell$.

Soit $\mathbb{A}^f$ l'anneau des adèles finis, produit restreint des $\mathbb{Q}_\ell$. On sait que pour tout sous-groupe ouvert K de $G(\mathbb{A}^f)$, l'ensemble des doubles classes $K\backslash G(\mathbb{A}^f)/G(\mathbb{Q})$ est fini (A. Borel, Some finiteness properties of adèle groups over number fields. Publ. Math. IHES 16 (1963), p. 5-30 -théorème 5.1). Cette finitude implique que $G(\mathbb{Q})$ n'a qu'un nombre fini d'orbites dans l'ensemble des réseaux $H' \subset H_{\mathbb{Q}}$ stables sous A_1. Tout réseau A-stable H' est contenu dans un réseau A_1-stable H'' avec un indice $[H'':H']$ qui divise $[A_1:A]^N$: prendre $H'' = AH'$, et on conclut en observant qu'un réseau n'a qu'un nombre fini de sous-réseaux d'indice donné.

3. Relation avec G. Faltings [4].

3.1. Le presént article a été inspiré par la lecture de G. Faltings [4].

On peut regarder [4] comme étant une autre preuve du théorème 0.1 dans le cas particulier où S est une courbe et où on ne considère que les schémas abéliens X sur S, d'une dimension relative fixe $n = N/2$, et la représentation de monodromie sur le H^1 des fibres. La restriction au cas des courbes est sans importance, car pour toute variété algébrique lisse S il existe une courbe lisse C tracée sur S telle que, pour $o \in C$, $\pi_1(C,o)$ s'envoie sur $\pi_1(S,o)$: pour U un ouvert dense de S, plongeable dans un espace projectif P, il suffit de prendre pour C l'intersection de U avec un sous-espace linéaire assez général de dimension dim P - dim S + 1; on aura $\pi_1(C) \twoheadrightarrow \pi_1(U)$ (Bertini) et $\pi_1(U) \twoheadrightarrow \pi_1(S)$.

Soient $\bar{S}$ une courbe projective et lisse, T un ensemble fini de points de $\bar{S}$ et $S := \bar{S} - T$. Soient X un schéma abélien sur S dont on suppose qu'il se prolonge en un schéma semi-abélien $\bar{X}$ sur $\bar{S}$ (réduction semi-stable). Soient n la dimension relative de X sur S, e la section nulle et posons $\omega := e^*\Omega^n_{\bar{X}/\bar{S}}$. C'est le faisceau inversible sur $\bar{S}$ dont la fibre en $s \in \bar{S}$ est la puissance extérieure maximale du dual de l'algèbre de Lie de la fibre $\bar{X}_s$ de $\bar{X}/\bar{S}$. Dans [4], G. Faltings commence par borner le degré de ω, indépendamment de X. Si on simplifie son argument par une référence à S. Zucker [11], on obtient l'estimation suivante.

<u>Lemme 3.2.</u> <u>Avec les notations précédentes, si</u> $-\chi(S) \geq o$, on a

$$\deg \omega \leq \frac{1}{2} n \cdot (-\chi(S)).$$

<u>Dans cette formule,</u> $\chi(S)$ <u>est la caractéristique d'Euler-Poincaré topologique: pour</u> $\bar{S}$ <u>de genre</u> g, $-\chi(S) = 2g - 2 + |T|$.

<u>Preuve</u>. Soit H la variation de structures de Hodge sur S de fibres les $H^1(X_s)$ $(s \in S)$. Elle donne lieu à un système local complexe $H_{\mathbb{C}}$, et à un fibré vectoriel complexe $\mathcal{H}$ muni d'une connection ∇, dont $H_{\mathbb{C}}$ est le système local des sections horizontales. Sur $\mathcal{H}$, on dispose de la filtration de Hodge F, réduite ici à un sous-fibré $F^1(\mathcal{H})$ de $\mathcal{H}$. La fibre $H^0(X_s,\Omega^1)$ de $F^1(\mathcal{H})$ en s s'identifie au dual de l'algèbre de Lie de X_s.

L'hypothèse que X est à réduction semi-stable équivaut à l'unipotence de la monodromie locale de $H_{\mathbb{C}}$ en chaque $t \in T$. Soit $\mathcal{H}_{can}$ le prolongement canonique ([7], 5.2) du fibré vectoriel $\mathcal{H}$ à $\bar{S}$ et soit $F^i(\mathcal{H}_{can})$ le sous-fibré localement facteur direct de $\mathcal{H}_{can}$ qui prolonge $F^i\mathcal{H}$. Toute polarisation de X (donc de H) induit une dualité parfaite entre $\mathcal{H}/F^1\mathcal{H}$ et $F^1\mathcal{H}$, et cette dualité se prolonge en une dualite parfaite entre $\mathcal{H}_{can}/F^1\mathcal{H}_{can}$ et $F^1\mathcal{H}_{can}$. Nous admettrons de la théorie des modèles de Néron que

$$e^*\Omega^1_{\bar{X}/\bar{S}} = F^1\mathcal{H}_{can}.$$

Nous traiterons d'abord du cas où X/S est sans partie fixe. Ceci équivaut à $H^0(S,H_{\mathbb{C}}) = 0$ et implique que $H^i(S,H_{\mathbb{C}}) = 0$ pour $i \neq 1$, d'où.

$$\text{(3.2.1)} \quad \dim H^1(S,H_{\mathbb{C}}) = -\chi(S,H_{\mathbb{C}}) = -\text{rang}(H_{\mathbb{C}}).\chi(S) = -2n\chi(S).$$

La cohomologie $H^*(S,H)$ peut se calculer comme l'hypercohomologie, sur $\bar{S}$, du complexe de De Rham

$$H_{can} \xrightarrow{\nabla} \Omega^1_S(T) \otimes H_{can}.$$

Ce complexe, noté K, est filtré par les sous-complexes

$$F^iK := (F^iH_{can} \to \Omega^1_S(T) \otimes F^{i-1}H_{can})$$

et, d'après S. Zucker [11], §13, la suite spectrale correspondante dégénère en E_1. En particulier,

(3.2.2) $\dim E_1^{0,1} + \dim E_1^{2,-1} \leq \dim H^1(S,H)$.

On a $E_1^{2,-1} = H^0(\Omega^1_S(T) \otimes F^1H_{can})$ et $E_1^{0,1} = H^1(H_{can}/F^1H_{can})$, de dual de Serre $H^0(\Omega^1_S \otimes F^1H_{can})$. Si F^1H_{can} est de degré d, i.e. si d := deg(w), on a donc

$$\begin{aligned}\dim E_1^{0,1} + \dim E_1^{2,1} &\geq \chi(\Omega^1_S \otimes F^1H_{can}) + \chi(\Omega^1_S(T) \otimes F^1H_{can}) \\ &= [n(g-1)+d] + [n(g-1)+ |T|) + d] \\ &= -n\chi(S) + 2d.\end{aligned}$$

Combiné avec (3.2.1) et (3.2.2), ceci donne

$$-n\chi(S) + 2d \leq -2n\chi(S),$$

i.e. l'assertion de 3.2.

Un schéma abélien X sur S est isogène à la somme d'un schéma abélien constant $S \times B$ et d'un schéma abélien Y sans partie fixe. Si X est à réduction semi-stable, Y l'est aussi. On a

$$\omega \text{ pour } X = (\omega \text{ pour } S \times B) \oplus (\omega \text{ pour } Y),$$

et le premier facteur est trivial. Ceci ramène 3.2 pour X à 3.2 pour Y, et termine la preuve de 3.2.

Remarque 3.3. Si $-\chi(S) \leq 0$, le groupe fondamental de S est abélien et tout schéma abélien sur S devient constant sur un revêtement fini étale de S. S'il a réduction semi-stable, il a bonne réduction, devient constant sur un revêtement fini étale de S et on a $\deg(\omega) = 0$.

Remarque 3.4. Dans le cas où X est une famille de courbes elliptiques (n=1), on dispose d'un morphisme "de Kodaira-Spencer"

(3.3.1) $\omega^{\otimes 2} \to \Omega^1_S(T)$.

S'il est trivial, l'invariant modulaire j est constant et $\deg(\omega)=0$ (X est toujours supposé à réduction semi-stable-donc, en l'occurence, à bonne réduction). S'il est non trivial, son existence reprouve 3.2.:

elle implique que $2\deg(\omega) \leq \deg(\Omega^1_{\bar{S}}(T))$ et on a

$$-\chi(S) = \deg \Omega^1_{\bar{S}}(T).$$

Pour X une famille modulaire, (3.3.1) est un isomorphisme et on a égalité dans (3.2).

Remarque 3.5. Supposons que $\chi(S) < 0$, i.e. que le revêtement universel de S est le disque unité D. C'est le cas intéressant (cf. 3.3). Une polarisation du schéma abélien X/S définit une polarisation de la variation de structures de Hodge H (H^1 des fibres), et en particulier une métrique hermitienne sur le fibré vectoriel $F^1\mathcal{H} = e^*\Omega^1_{X/S}$. La variation H donne lieu à une application de périodes $P : D \to M$ comme en 1.7. Le théorème de Griffiths 1.7 que P décroît les distances entraîne une majoration du tenseur de courbure du fibré métrisé $e^*\Omega^1_{X/S}$, et en particulier de sa 2-forme de Chern. Une étude locale à l'infini montre que l'intégrale de cette 2-forme est le degré de ω sur $\bar{S}$ (C. Peters [12]). Intégrant la majoration ci-dessus, on retrouve 3.2.

Remarque 3.6. Soient H une variation complexe de structures de Hodge polarisable sur une courbe $S = \bar{S} - T$, $(\mathcal{H},\nabla)$ le fibré vectoriel à connection correspondant, $\mathcal{H}_{can}$ le prolongement canonique de $\mathcal{H}$, et F la filtration de Hodge de $\mathcal{H}_{can}$: $F^i\mathcal{H}_{can}$ est le sous-fibré localement facteur direct de $\mathcal{H}_{can}$ qui prolonge le sous-fibré $F^i\mathcal{H}$ de $\mathcal{H}$. Le théorème de Griffiths 1.7 et une étude locale à l'infini ([12]) permettent encore de majorer les $|\deg \mathrm{Gr}^i_F \mathcal{H}_{can}|$ en fonction du genre de S, de $|T|$ et des nombres de Hodge de la variation H.

3.7. Soient n un entier ≥ 3 et $M_{d,n}$ l'espace des modules des variétés abéliennes principalement polarisées de dimension d munies d'une structure de niveau n. C'est un espace fin de modules: la donnée d'un schéma abélien principalement polarisé X S, de dimension relative d, muni d'une structure de niveau n, équivaut à celle d'un morphisme $f : S \to M_{d,n}$. Pour $S = \bar{S} - T$ comme en 3.1, ce morphisme se prolonge en un morphisme $\bar{f}$ de $\bar{S}$ dans la compactification de Satake $\bar{M}_{d,n}$ de $M_{d,n}$. Faltings déduit de 3.2 que le graphe $\Gamma(\bar{f}) \subset \bar{S} \times \bar{M}_{d,n}$ de $\bar{f}$ est de degré borné indépendamment de X/S, pour un plongement projectif convenable de $\bar{S} \times \bar{M}_{d,n}$. De là résulte que les schémas abéliens du type considéré sont paramétrés par un schéma de type fini Y: il existe un schéma abélien X_Y sur $S \times Y$ tel que chaque schéma abélien du type dit sur S soit isomorphe à la restriction X_y de

X_Y à $S \times \{y\} \sim S$. La monodromie étant un invariant discret, la classe d'isomorphie de la représentation de monodromie de $\pi_1(S)$ définie par X_y ne dépend que de la composante connexe de Y où se trouve y. Ceci reprouve le théorème 0.1, restreint au cas particulier des schémas abéliens de dimension relative d, principalement polarisés et munis d'une structure de niveau n.

3.8. Soient Z un schéma connexe de type fini et $u : S_Z \to Z$ une famille de courbes paramétrisée par Z. On suppose que $S_Z = \bar{S}_Z - T_Z$ avec $\bar{S}_Z$ propre et lisse sur Z, de fibres des courbes de genre g, et avec T_Z fini étale sur Z, de fibres des ensembles de t points. Pour $z \in Z$, posons $S_z := u^{-1}(z)$. Pour $x \in S_Z$, les groupes fondamentaux $\pi_1(S_{u(x)},x)$ des fibres de u forment un système local sur S_Z. Un chemin de x à y définit donc un isomorphisme de $\pi_1(S_{u(x)},x)$ à $\pi_1(S_{u(y)},y)$. Si, pour $i = 1,2$, S_i est le complément dans une courbe $\bar{S}_i$ de genre g d'un ensemble T_i de t points, et que $x_i \in S_i$, un isomorphisme de $\pi_1(S_1,x_1)$ avec $\pi_1(S_2,x_2)$ sera dit <u>permis</u> s'il existe une famille comme ci-dessus, dont les S_i soient des fibres et pour laquelle l'isomorphisme soit déduit d'un chemin de x_1 à x_2. L'espace de modules des courbes de genre g étant connexe, il existe toujours un isomorphisme permis de $\pi_1(S_1,x_1)$ avec $\pi_1(S_2,x_2)$.

3.9. Les arguments 3.6 s'appliquent encore avec paramètres. Pour une famille $u : S_Z \to Z$ comme en 3.7 et $d \geq 0$, $n \geq 3$, il existe un schéma de type fini Y sur Z et un schéma abélien X sur $S_Y := S_Z \times_Z Y$ ayant la propriété suivante: pour tout $z \in Z$ et tout schéma abélien principalement polarisé de dimension d, A, sur S_z, muni d'une structure de niveau n, il existe $y \in Y$ au-dessus de z tel que A soit l'image inverse de X par $S_z \to S_Y : s \to (s,y)$. Prenant une famille universelle S_Z, on en déduit le résultat suivant.

<u>3.10. Proposition</u>. <u>Soit</u> Γ <u>le groupe fondamental d'une courbe de genre</u> g <u>avec</u> t <u>points ôtés. Fixons et</u> $n \geq 3$. <u>Pour toute courbe de genre</u> g <u>avec</u> t <u>points ôtés</u> S, <u>pour tout schéma abélien</u> A <u>sur</u> S, <u>principalement polarisé, de dimension</u> d, <u>muni d'une structure de niveau</u> n, <u>et pour tout isomorphisme permis</u> γ <u>de</u> Γ <u>avec</u> $\pi_1(S,o)$, <u>la représentation de monodromie</u> σ <u>de</u> $\pi_1(S,o)$ <u>sur</u> $H^1(A_o,\mathbb{Z})$ <u>fournit une représentation</u> $\sigma\gamma$ <u>de</u> Γ. <u>Les classes</u>

d'isomorphie de représentations de Γ ainsi obtenues ne forment qu'un nombre fini d'orbites sous les automorphismes permis de Γ.

3.11 Remarque. L'analogue de 3.9, sans structure de niveau imposée, est encore vrai. Tout schéma abélien de dimension d sur une courbe S acquiert une structure de niveau 3 sur un revêtement de S de degré $< 3^{4d^2}$, et on décrit un schéma abélien sur S par la donnée d'un revêtement S' de S, de degré $< 3^{4d^2}$, d'un schéma abélien avec structure de niveau 3 sur S', et d'une donnée de descente de S' à S pour ce schéma abélien.

3.12 Remarque. On en déduit que l'analogue de 3.10, sans structure de niveau imposée, est encore vrai. On peut aussi se passer des polarisations en utilisant le théorème de Zarhin que pour tout schéma abélien polarisable A, il existe une polarisation principale sur $(A \times A^{dual})^4$ (voir par exemple Astérisque 127 (séminaire sur les pinceaux arithmétiques: la conjecture de Mordell, dirigé par L. Szpiro) VII 1.)

3.13 Question. Si on remplace "schéma abélien" par "variation de structures de Hodge", l'énoncé 3.10 reste-t-il valable?

Bibliographie

[1] P. Deligne, Equations différentielles à points singuliers réguliers, Lecture Notes in Mathematics 163, Springer-Verlag 1970.

[2] P. Deligne, Théorie de Hodge II, Publ. Math. IHES 40 (1971), 5-58.

[3] P. Deligne and D. Mostow, Monodromy of hypergeometric functions and non-lattice integral monodromy, Publ. Math. IHES, to appear (1986).

[4] G. Faltings, Arakelov's theorem for abelian varieties, Inv. Math. 73 (1983), 337-348.

[5] P.A. Griffiths, Periods of integrals on algebraic manifolds: summary of main results and discussion of open problems, Bull. AMS 76 2 (1970), 228-296.

[6] P.A. Griffiths, Periods of integrals on algebraic manifolds III, Publ. Math. IHES 38 (1970), 125-180.

[7] P.A. Griffiths and W. Schmid, Recent developments in Hodge theory: a discussion of techniques and results, In: discrete subgroups of Lie groups and applications to moduli, TIFR 1973, Oxford University Press 1975.

[8] J. Morgan, The algebraic topology of smooth algebraic varieties, Publ. Math. IHES 48 (1978), 137-204.

[9] M. Nori, Non arithmetic monodromy, C.R. Acad. Sci. Paris, to appear (1986).

[10] W. Schmid, Variations of Hodge structures: the singularities of the period mapping, Inv. Math. 22 (1973), 211-319.

[11] S. Zucker, Hodge theory with degenerating coefficients, Ann. of Math. 109 (1979), 415-476.

[12] C. Peters, A criterion for flatness of Hodge bundles and geometric applications, Math. Ann. 268 (1984), 1-19.

[13] C. Procesi, The invariant theory of n × n matrices, Adv. in Math. 19 (1976), 306-381.

SCHOOL OF MATHEMATICS
THE INSTITUTE FOR ADVANCED STUDY
PRINCETON, NEW JERSEY 08540

SOME ASPECTS OF THE ARITHMETIC THEORY OF POLYNOMIALS*

by Jun-ichi Igusa

Introduction

This is an expository paper based on our memos of two lectures, one at the A.M.S. annual meeting at Cincinnati in January of 1982 with Professor Mostow presiding and another at the conference at Yale in honor of his 60th birthday. We have emphasized the universality or the uniformity of results and problems for all local fields; and consequently we have included certain material which is usually considered as analysis rather than arithmetic. On the other hand, as the title suggests, we have covered only certain parts of the arithmetic theory of polynomials. For instance the arithmetic theory of polynomials over a finite field is almost entirely left out. Also we have not given enough explanation to the recent results of Barlet [3] and Heath-Brown [11], which were mentioned at the time of the conference respectively by Professors Deligne and Tamagawa. Nevertheless it is our hope that this paper will give a fair and useful survey of certain parts of the arithmetic theory of polynomials appropriate for the Festschrift.

§1. One variable case

The arithmetic theory of an arbitrary polynomial is a generalization of that of the polynomial x. In order to make this point clearer later on and also to fix our notation we shall start by recalling some well-known theorems on the polynomial x. We refer to Tate [47] and Weil [50] for the details.

In general if R is any associative ring with 1, we shall denote by $R^{\times}$ the group of units of R. For any locally compact group G we

*This work was partially supported by the National Science Foundation.

shall denote by $\Omega(G)$ the topological group of all continuous homomorphisms $\omega: G \to \mathbb{C}^{\times}$. If G is abelian, we shall denote by $S(G)$ the Schwartz-Bruhat space of G and by $S(G)'$ the space of tempered distributions in G.

We shall denote by k a number field, i.e., a finite algebraic extension of $\mathbb{Q}$, by k_v its completion with respect to a normalized absolute value $||_v$ on k, and by A_k the adele ring of k, which is a restricted product of all k_v. If X is a k-variety, we shall denote $X(k)$, $X(k_v)$, $X(A_k)$ by X_k, X_v, X_A in that order; X_k is, therefore, the set of k-rational points of X. We recall that X_v, X_A are locally compact spaces and that if X is affine, then X_k is discrete in X_A. In particular if $X = \mathrm{Aff}^n$, the affine n-space, then $X_k = k^n$ is discrete in $X_A = A_k^n$ with compact quotient. If G is an algebraic k-group, then G_A is a locally compact group; in particular $I_k = (GL_1)_A$ is the idele group of k. If $x = (x_v)_v$ is in I_k, then $|x|_A = \prod_v |x_v|_v$ is defined and it takes the value 1 on $k^{\times}$. Furthermore if I_k^1 denotes the subgroup of I_k defined by $|x|_A = 1$, then $k^{\times}$ is discrete in I_k, $I_k^1/k^{\times}$ is compact, and $||_A : I_k/I_k^1 \cong R_+^{\times}$, the multiplicative group of positive real numbers.

We observe that the identity component of $\Omega(I_k/k^{\times})$ consists of $\omega_s(x) = |x|_A^s$ for all s in $\mathbb{C}$; that for every ω in $\Omega(I_k/k^{\times})$ there exists a unique real number $\sigma(\omega)$ satisfying $|\omega(x)| = \omega_{\sigma(\omega)}(x)$. For any real number σ we shall denote by $\Omega_\sigma(I_k/k^{\times})$ the open subset of $\Omega(I_k/k^{\times})$ defined by $\sigma(\omega) > \sigma$. The situation is similar for $\Omega(k_v^{\times})$ and we shall use the same notation, such as $\Omega_\sigma(k_v^{\times})$, as above.

We shall recall a Poisson formula and the functional equations of ω both local and global: we take an affine space X defined over k such that X_k is equipped with a nondegenerate k-bilinear form $[x,y]$; we shall denote by $|dx|_A$ the Haar measure on X_A normalized by the condition that the volume of X_A/X_k is 1; and we define a Fourier transformation $\Phi \to \Phi^*$ in $S(X_A)$ relative to an element $\psi \neq 1$ of $\Omega(A_k/k)$ as

$$\Phi^*(x) = \int_{X_A} \psi([x,y])\Phi(y)\ |dy|_A .$$

If g is a bicontinuous automorphism of X_A and if $[x,gy] = [{}^t gx,y]$, $|d(gx)|_A = |g|_A\ |dx|_A$, then the Poisson formula states that

$$\text{(P)} \quad \sum_{\xi \in X_k} \Phi(g\xi) = |g|_A^{-1} \cdot \sum_{\xi \in X_k} \Phi^*({}^t g^{-1}\xi),$$

in which both sides are absolutely convergent. We might mention that the generalization from the special case where $g = 1$ to the above case is straightforward; and that if $\mathrm{End}(X)$ denotes the ring of linear transformations in X and if $GL(X) = \mathrm{End}(X)^\times$, then for any g in $GL(X)_A$ we have $|g|_A = |\det(g)|_A$.

We take a Haar measure μ on I_k and for any ω in $\Omega_1(I_k/k^\times)$ and any Φ in $S(A_k)$ we put

$$\text{(ZG1)} \quad T_\omega(\Phi) = \int_{I_k} \omega(x)\Phi(x)\mu(x);$$

the integral is absolutely convergent and it defines a holomorphic function $\omega \to T_\omega(\Phi)$ on $\Omega_1(I_k/k^\times)$. A fundamental theorem states that this holomorphic function has a meromorphic continuation to the whole $\Omega(I_k/k^\times)$ with poles of order 1 at ω_0, ω_1 and satisfies the functional equation $T_\omega(\Phi^*) = T_{\omega_1\omega^{-1}}(\Phi)$, in which the Fourier transformation is relative to $[x,y] = xy$ for $X = \mathrm{Aff}^1$. If μ is symbolically the product of the Haar measure on I_k^1 normalized by $\mathrm{vol}(I_k^1/k^\times) = 1$ and $d\log|x|_A$ on I_k/I_k^1, the respective residues are $-\Phi(0)$, $\Phi^*(0)$. In an oversimplified manner these are obtained by applying the Poisson formula (P) to the integrand of

$$\text{(ZG2)} \quad T_\omega(\Phi) = \int_{I_k/k^\times} \Big(\sum_{\xi \in k^\times} \Phi(x\xi)\Big)\omega(x)\mu(x)$$

with x as "g". We define T^* for any T in $S(A_k)'$ as $T^*(\Phi) = T(\Phi^*)$ and we say that T_ω is an $S(A_k)'$-valued meromorphic function on $\Omega(I_k/k^\times)$ satisfying the functional equation

$$\text{(FG)} \quad T^*_\omega = T_{\omega_1\omega^{-1}}.$$

If we put $K = k_v$ and $|\ |_K = |\ |_v$, the canonical injection $k_v \to A_k$ followed by ψ defines an element ψ_v of $\Omega(K)$. If $K = \mathbb{R}$, $\mathbb{C}$, we define ψ_K as $\psi_K(x) = e(x)$ $(= \exp(2\pi i x))$, $e(2\cdot \mathrm{Re}(x))$; if K is a p-adic field with 0_K as its ring of integers and $\pi 0_K = 0_K - 0_K^\times$, we take any ψ_K from $\Omega(K)$ with the property that $\psi_K = 1$ on 0_K but not on $\pi^{-1}0_K$. Then we can write

$\psi_v(x) = \psi_K(a_v^{-1}x)$ with a_v in $k_v^\times$. Furthermore $a = (a_v)_v$ is in I_k and $|a|_A = |D|$, where D is the discriminant of k. On the other hand we shall denote by $|dx|_K$ the Haar measure on K, and later its product-measure on K^n, normalized by $\mathrm{vol}\{x \in K; |x|_K \leq 1\} = 2,\ 2\pi$, or 1 according as $K = \mathbb{R},\ \mathbb{C}$, or a p-adic field. If we denote $|a_v|_v^{-1/2}\ |dx|_K$ by $|dx|_v$, the restricted-product measure of all $|dx|_v$ becomes $|dx|_A$. The Haar measures $|dx|_K$, $|dx|_v$, $|dx|_A$ are the self-dual measures on K, $k_v = K$, A_k relative to ψ_K, ψ_v, ψ in that order.

Now every ω in $\Omega_1(K^\times)$ defines an element T_ω of $S(K)'$ as

$$\text{(ZL)}\quad T_\omega(\Phi) = \int_{K^\times} \omega(x)\Phi(x)\cdot|x|_K^{-1}|dx|_K.$$

This time we define Fourier transformations in $S(K)$, $S(K)'$ relative to $\psi_K(xy)$ and $|dx|_K$. Then T_ω becomes an $S(K)'$-valued meromorphic function on $\Omega(K^\times)$ and satisfies a functional equation of the form

$$\text{(FL)}\quad T_\omega^* = \Gamma(\omega)T_{\omega_1\omega^{-1}}$$

with a $\mathbb{C}$-valued meromorphic function $\Gamma(\omega)$ on $\Omega(K^\times)$.

If we start from ω in $\Omega(I_k/k^\times)$ and define ω_v as the canonical injection $k_v^\times \to I_k$ followed by ω, the functional equations (FG) and (FL) for ω and ω_v imply

$$\prod_v \omega_v(a_v)\Gamma(\omega_v) = |D|^{1/2},$$

where the infinite product is taken separately for the numerator and the denominator of $\Gamma(\omega_v)$; this is the functional equation for the Hecke L-series associated with ω.

§2. Siegel-Weil formula

We shall consider a triplet (G,X,ρ), in which G is a reductive algebraic group, X is an affine space, and ρ is a rational representation of G in X all defined over a number field k; we shall exclude the trivial case where $\rho(g)x = x$ for every g in G. The one variable case is the theory of $(GL_1, \mathrm{Aff}^1, \rho)$, in which $\rho(g)x = gx$. We shall explain some aspects of the line of investigation:

"Siegel's main theorem" → "Siegel-Weil formula" → ... .

We shall assume that G is connected and semisimple. We can then normalize a Haar measure μ on G_A by $vol(G_A/G_k) = 1$, cf. Borel [6]; and for any Φ in $S(X_A)$ we put

$$I(\Phi) = \int_{G_A/G_k} \Big(\sum_{\xi \varepsilon X_k} \Phi(\rho(g)\xi)\Big)\mu(g).$$

The integrand is a continuous function on G_A/G_k, but in general it is not in $L^1(G_A/G_k)$. In fact all triplets such that similar integrals relative to algebraic extensions of k remain convergent are so special that they have been classified; and by using the classification and the theory of algebras the following theorem has been proved:

"If a triplet (G,X,ρ) satisfies the above condition of convergence, the ring of invariants of ρ is generated by algebraically independent homogeneous elements, say $f_1, \ldots, f_r$, of the symmetric algebra of X_k; if we define a k-morphism $f:X \to Aff^r$ as $f(x) = (f_1(x), \ldots, f_r(x))$, then X contains a k-open subset X' such that at every ξ in X' the r hypersurfaces

$$f_1(x) - f_1(\xi) = 0, \ldots, f_r(x) - f_r(\xi) = 0$$

are transversal and such that $U(i) = f^{-1}(i) \cap X'$ is a G-orbit with the codimension of $f^{-1}(i) - U(i)$ in $f^{-1}(i)$ at least 2 for every i in Aff^r. Furthermore the fixer H_ξ of ξ in G is connected and its radical is unipotent; if G is simply connected and if ρ does not contain the third fundamental representation of Sp_6 as an irreducible constituent, H_ξ is also simply connected, i.e., its semisimple part is simply connected."

We refer to [13] for the classification and to [14] for a proof of the above theorem. We put Aff^r into a duality with itself via $[i,i^*] = i_1i_1^* + \ldots + i_ri_r^*$; then the above theorem permits us to formulate (the main part of) a conjectural Siegel-Weil formula as follows:

"If ρ does not contain the third fundamental representation of Sp_6 as an irreducible constituent, then

(SW) $\int_{G_A/G_k} \left(\sum_{\xi \varepsilon X_k'} \Phi(\rho(g)\xi) \right) \mu(g) = \sum_{i^* \varepsilon k^r} \int_{X_A} \psi([i^*, f(x)]) \Phi(x) \, |dx|_A$;

the series on the R.H.S. is absolutely convergent for every Φ in $S(X_A)$."

We shall recall a short history after Siegel: in late 1950's Tamagawa discovered that Siegel's main theorem on quadratic forms was equivalent to an intrinsically defined volume of any special orthogonal group being 2; cf. [46]. This viewpoint was further developed by Weil [48] and later the analytic formulation of Siegel's main theorem was given the form of the Siegel formula by him [49], which is now called the Siegel-Weil formula. In doing so important generalizations as well as simplifications were made with the famous conjecture stating that the Tamagawa number of any connected simply connected semisimple algebraic group is 1 becoming more certain. We shall briefly recall the definition of Tamagawa numbers and clarify the relation between (SW) and the above conjecture by Weil.

If U is any irreducible smooth k-variety, an everywhere regular nowhere zero $\dim(U)$-form θ on U defined over k is called a gauge form on U. If such a θ exists, it gives rise to a positive measure $|\theta|_v$ on U_v based on the measure $|dx|_v$ on k_v in §1. The restricted-product measure $|\theta|_A$ of all $|\theta|_v$, if it exists, is called a Tamagawa measure on U_A. If G is, for a moment, any connected algebraic group defined over k, a gauge form dg exists; and if the radical of G is unipotent, a Tamagawa measure $|dg|_A$ on G_A exists and is intrinsic, i.e., it is independent of the choice of dg. Furthermore by an already quoted theorem of Borel

$$\tau_k(G) = \int_{G_A/G_k} |dg|_A$$

is finite; this intrinsic volume of G_A/G_k is the Tamagawa number of G.

We shall express (SW) as $I'(\Phi) = E'(\Phi)$; we observe that $I'(\Phi)$ is only a part of $I(\Phi)$. As we shall explain by an example in §3, the Siegel-Weil formula is obtained by completing $I'(\Phi) = E'(\Phi)$. If $\tilde{G}$ is a simply connected covering group of G defined over k, we have $I'(\tilde{G}) = I'(G)$. This was proved by Haris [10] using Ono's formula for the Tamagawa numbers of isogenous connected semisimple algebraic groups; cf. [33]. Therefore we may assume that G is simply connected if that is convenient.

Now for every i in k^r the quotient of gauge forms, say dx and di, on X and Aff^r gives a gauge form $\theta_i(x) = (dx/df(x))_i$ on $U(i)$; and it gives rise to an intrinsic Tamagawa measure $|\theta_i|_A$ on $U(i)_A$. In addition to $I'(\Phi)$ and $E'(\Phi)$ we introduce

$$I''(\Phi) = \sum_{i \varepsilon k^r} \int_{U(i)_A} \Phi(x) \, |\theta_i(x)|_A .$$

Then modulo "Hasse principles," one of which will be explained in §3, (SW) will follow from, in fact it is almost equivalent to, the combination of

(T) $\tau_k(G) = \tau_k(H_\xi)$ for every ξ in X_k'

($P^{\#}$) $I''(\Phi) = E'(\Phi)$ for every Φ in $\mathcal{S}(X_A)$.

We recall that if G is simply connected, so is H_ξ in the sense explained in our theorem. Therefore if Weil's conjecture is true, we have $\tau_k(G) = \tau_k(H_\xi) = 1$. The point is that (T) gives the possibility of proving $\tau_k(G) = 1$ inductively because the semisimple part of H_ξ is a smaller group than G. As for ($P^{\#}$), if we consider the trivial case, which we have excluded, where $\rho(g)x = x$ for every g in G, it reduces to the Poisson formula (P) in §1; hence ($P^{\#}$) is a generalized Poisson formula.

The present state of affairs of (SW) is as follows: firstly the general case can be reduced via the classification to the case where G is absolutely simple; secondly once (SW) is proved, to complete it into a Siegel-Weil formula has been easy. And it has been proved for classical representations of classical groups by Weil [49] except for one case later proved by Mars [29]. Also it has been proved in the cases where $r = 1$ except for the 56-dimensional irreducible representation defined over k of a connected simply connected simple group of type E_7; in that case ($P^{\#}$) but not (T) has been proved; cf. [17]. Among these Mars' case [28] is the first case where a Siegel-Weil formula involving an invariant of higher degree has been proved; there ρ is the 27-dimensional irreducible representation defined over k of a connected simply connected simple group of type E_6 and $f(x)$ is a cubic invariant.

§3. Metaplectic and hypermetaplectic groups

The reason why we have difficulty in proving (SW) for a certain k-form of type E_7 is that the inductive approach explained in §2 involves infinitely many k-forms of type E_6 with unknown Tamagawa numbers each having invariants of degrees 2, 3, 3, 4; and that at the present moment we only have the following criterion for the validity of $(P^{\#})$:

"Let $f(x)$ denote any polynomial in n variables $x_1, \dots, x_n$ with coefficients in a number field k and define a k-closed subset C_f of Aff^n, called its critical set, by

$$\partial f/\partial x_1 = \dots = \partial f/\partial x_n = 0;$$

assume that C_f is a subset of $f^{-1}(0)$ of codimension at least 2 and that there exists a fixed $\sigma > 2$ satisfying

$$\left|\int_{O_K^n} \psi_K(i^*f(x))\,|dx|_K\right| \leq \max(1, |i^*|_K)^{-\sigma}$$

for every i^* in $K = k_v$ and for almost all v; then $(P^{\#})$ holds:

$$\sum_{i \in k} \int_{U(i)_A} \Phi(x)\ |\theta_i(x)|_A = \sum_{i^* \in k} \int_{X_A} \psi(i^*f(x))\Phi(x)\ |dx|_A$$

with the understanding that $U(i) = f^{-1}(i)$ for $i \neq 0$ and $U(0) = f^{-1}(0) - C_f$."

This criterion is usually applied to the case where $f(x)$ is homogeneous; a proof in that case is in [15], Lect. Notes. Later Yamazaki has pointed out that the proof, after a modification, becomes valid for the above improved criterion. At any rate if $f(x)$ is a nondegenerate quadratic form in $n > 4$ variables, the conditions can easily be verified; and if we take $SO(f)$ as G, Aff^n as X, and put $\rho(g)x = gx$, we can prove $\tau_k(G) = 2$ and the Siegel-Weil formula in an over-simplified manner as follows:

If we define Fourier transformations in $\mathcal{S}(X_A)$ and $\mathcal{S}(X_A)'$ relative to $[x,y] = f(x + y) - f(x) - f(y)$, we have $I(\Phi^*) = I(\Phi)$ and $\psi(i^*f(x))^* = \psi(-(i^*)^{-1}f(x))$ for every i^* in $k^\times$. On the other hand by an induction on n and by using $(P^{\#})$ we get $I(\Phi) = \Phi(0) + (2/\tau_k(G))E'(\Phi)$. By putting these together we see that

$\Phi(0) + (2/\tau_k(G))\Phi^*(0)$ is invariant under $\Phi \to \Phi^*$ for every Φ in $S(X_A)$, hence $\tau_k(G) = 2$. If we put $E(\Phi) = \Phi(0) + E'(\Phi)$, we can write

$$I(\Phi) = E(\Phi),$$

which is the Siegel-Weil formula in this case.

In the above proof of $I(\Phi) = E(\Phi)$ we have tacitly used the Hasse principle stating that if $U(i)_k$ is empty, so is $U(i)_A$ for every i in k. In [49] Weil has characterized E as an element of $S(X_A)'$ and derived $I = E$ as its consequence; this proof implies the Hasse principle. In his characterization of E he has used a metaplectic group; in view of its great importance we shall at least recall its definition:

In the general case by using the classification we can show that there exists a nondegenerate k-bilinear form $[x,y]$ on X_k such that ${}^t\rho(G) = \rho(G)$; we define a Fourier transformation in $S(X_A)$ relative to $[x,y]$. We denote by $Mp(f)_A$ (resp. $Mp(f)_k$) the subgroup of the unitary group of $L^2(X_A)$ generated by the multiplication by $\psi([i^*,f(x)])$ as i^* runs over A_k^r (resp. k^r) and by its conjugate under the Fourier transformation; then $s \to I(s\Phi)$ becomes a function on $Mp(f)_k\backslash Mp(f)_A$, i.e., a function on $Mp(f)_A$ which is automorphic under $Mp(f)_k$. If ρ is a classical representation of a classical group, then $Mp(f)_A$ becomes the global metaplectic group of Weil; actually he has included in his $Mp(f)_A$ the group of scalar multiplications by complex numbers of absolute value 1.

We go back to the case where $f(x)$ is a nondegenerate quadratic form: in that case there exists a continuous homomorphism $Mp(f)_A \to (SL_2)_A$, which is proper, hence surjective, and under which $Mp(f)_k$ is mapped isomorphically to $(SL_2)_k$. Actually the kernel of $Mp(f)_A \to (SL_2)_A$ is of order 1 or 2 according as n is even or odd; but this information is not necessary to conclude that $Mp(f)_k\backslash Mp(f)_A$ has only one cusp. The point is that by using this fact and $(P^{\#})$ only, i.e., without using the Siegel-Weil formula, we can prove the following asymptotic relation:

$$\sum_{\xi\in U(0)_k} \Phi(t\xi) = \int_{U(0)_A} \Phi(x)\ |\theta_0(x)|_A \cdot |t|_A^{2-n} + o(|t|_A^{2-n})$$

as $|t|_A \to 0$ in I_k; this implies the Hasse principle. A detailed account for these is, e.g., in Ariturk [1].

If more generally $f(x)$ is a form, i.e., a homogeneous polynomial of degree $m \geq 2$ in $n > 2m$ variables with coefficients in k such that $C_f = \{0\}$, we can verify the conditions for the validity of $(P^{\#})$. This time, however, the verification requires a deep theorem of Deligne [8]. At any rate the well-known theorem stating that the behavior of a theta series at a cusp can be described up to lower order terms by an Eisenstein series has been generalized. In order to prove the Hasse principle along this line it is desirable to have a reduction theory for $Mp(f)_A$. With that in mind we have examined a local metaplectic group over $\mathbb{R}$ and obtained the following result:

"In general let $f(x)$ denote a homogeneous polynomial of degree $m \geq 1$ in $n \geq 1$ variables $x_1, \ldots, x_n$ with coefficients in $\mathbb{R}$ and $\Phi \to \Phi^*$ the Fourier transformation in $\mathcal{S}(\mathbb{R}^n)$ relative to $[x,y] = x_1y_1 + \ldots + x_ny_n$; then the subgroup $Mp(f)_{\mathbb{R}}$ of the unitary group of $L^2(\mathbb{R}^n)$ generated by the multiplication by $\psi_{\mathbb{R}}(tf(x))$ as t runs over $\mathbb{R}$ and by its conjugate under the Fourier transformation is finite dimensional if and only if $m = 1, 2$."

The tangent vectors at $t = 0$ of the above one-parameter subgroups are, up to constant factors, the multiplication by $f(x)$ and the application of $f(\partial/\partial x)$; and the Lie algebra generated by $f(x)$ and $f(\partial/\partial x)$ is shown in [18] to be infinite dimensional if $m > 2$. We have tried to save the finite dimensionality and classified all finite dimensional Lie algebras generated by an element $f(x)$ of $\mathbb{R}[x]$ and an element of $\mathbb{R}[x,d/dx]$; and we have found rather disappointingly that all finite dimensional Lie algebras in which $\deg(f(x)) > 2$ are solvable; cf., "Some observations on metaplectic groups," Amer. J. Math. 103 (1981), pp. 1343-1365.

In view of the above circumstances it appears more appropriate to call any infinite dimensional "Mp" a hypermetaplectic group. The local hypermetaplectic groups over $\mathbb{R}$ or $\mathbb{C}$ are not the kind of infinite dimensional Lie groups appearing in the works of Lie and Cartan. Professor Kac has told us during a recent conference in Rome that they are very likely related to the kind of infinite dimensional Lie groups appearing in his works; cf. [21].

Finally Heath-Brown [11] has succeeded in proving that $U(0)_k$ is not empty if $f(x)$ is a form of degree $m = 3$ in $n > m^2 = 9$

variables with coefficients in $k = \mathbb{Q}$ such that $C_f = \{0\}$. It is hoped that an interpretation of his work in adelic language eventually allows one to generalize his theorem to the case where k and m are arbitrary.

§4. Uniform theory of functions F_Φ, F^*_Φ, Z_Φ

We shall now explain a local theory by which the "criterion" in §3 has been proved: we shall denote by K any completion of a number field as in §1 and by $f(x)$ a polynomial in n variables $x_1, \ldots, x_n$ with coefficients in K. We shall exclude the trivial case where the map $f:K^n \to K$ is constant and denote by C_f its critical cet; C_f is a closed subset of K^n and by a theorem of Bertini $f(C_f)$ is finite. For instance if $f(x)$ is homogeneous of degree m, then $f(C_f)$ is empty or $\{0\}$ according as $m = 1$ or $m \geq 2$.

If Φ is in $S(K^n)$ and i^* is in K, then

$$F^*_\Phi(i^*) = \int_{K^n} \Phi(x)\psi_K(i^*f(x))\ |dx|_K$$

defines a bounded uniformly continuous function F^*_Φ on K; if ω is in $\Omega_0(K^\times)$, then

$$Z_\Phi(\omega) = \int_{K^n} \Phi(x)\omega(f(x))\ |dx|_K$$

defines a holomorphic function Z_Φ on $\Omega_0(K^\times)$. These two functions are related via a continuous function F_Φ on $K - f(C_f)$ contained in $L^1(K)$: if i is in $K - f(C_f)$ and $dx = dx_1 \wedge \ldots \wedge dx_n$, then $\theta_i(x) = (dx/df(x))_i$ gives rise to a positive measure $|\theta_i|_K$ on $f^{-1}(i)$ based on the measure $|dx|_K$; the function F_Φ is defined as

$$F_\Phi(i) = \int_{f^{-1}(i)} \Phi(x)\ |\theta_i(x)|_K$$

and the relations are

$$F^*_\Phi(i^*) = \int_K F_\Phi(i)\psi_K(ii^*)\ |di|_K,\quad Z_\Phi(\omega) = \int_K F_\Phi(i)\omega(i)\ |di|_K\ .$$

By lifting a partition of unity on K to K^n and by translations in

K the general case can be reduced to the case where $\mathrm{Supp}(\Phi) \cap C_f \subset f^{-1}(0)$; then the following objects are related by formulas:

(i) Principal parts of Laurent expansions of the meromorphic continuation of $Z_\Phi(\omega)$ around its poles;

(ii) Terms of asymptotic expansions of $F_\Phi(i)$ as $|i|_K \to 0$;

(iii) Terms of asymptotic expansions of $F^*_\Phi(i^*)$ as $|i^*|_K \to \infty$.

The asymptotic expansions in (ii), (iii) are both in terms of certain ω's and their derivatives. As for (i), if we denote by U_K the subgroup of $K^\times$ defined by $|i|_K = 1$, we have the following basic finiteness: if K is a p-adic field, the set of $\omega|U_K$ for which $Z_\Phi(\omega)$ does not vanish identically is finite; if $K = \mathbb{R}, \mathbb{C}$, for any σ in $\mathbb{R}$ the set of $\omega|U_K$ for which $Z_\Phi(\omega)$ is not holomorphic on $\Omega_\sigma(K^\times)$ is finite. Furthermore in the p-adic case $Z_\Phi(\omega)$ for each $\omega|U_K$ is a rational function of $t = \omega(\pi)$. We recall that in the archimedean case, if we write $\omega(i) = |i|_K^s (i/|i|)^p$ with p in $\mathbb{Z}$, the poles of $Z_\Phi(\omega)$ in the s-plane are negative rational numbers.

We have inherited the notation F_Φ, F^*_Φ and some results from Weil [49]; these functions and Z_Φ were known in analysis at least in the case where $K = \mathbb{R}$. The fact that the meromorphic continuation of $Z_\Phi(\omega)$ can be proved by using Hironaka's theorem on desingularization was discovered by Bernshtein-Gel'fand [4] and Atiyah [2]; the asymptotic expansion of $F_\Phi(i)$ was proved along that line by Jeanquartier [20]. Later Malgrange [27] has examined the relation between the monodromy of f and the terms of asymptotic expansions of $F_\Phi(i)$, $F^*_\Phi(i^*)$. We have developed a uniform theory of F_Φ, F^*_Φ, Z_Φ valid for all K in [15]. For our later purpose we shall recall the way how the principal parts of the Laurent expansions of $Z_\Phi(\omega)$ are determined by the resolution data of the singularities of $f(x) = 0$.

We put $X = \mathrm{Aff}^n$, hence $X_K = K^n$, and in the present local situation we drop the subscript K from X_K; with this general notational agreement a consequence of Hironaka's theorem [12], p. 176 can be stated as follows: there exist an everywhere n-dimensional K-analytic manifold Y and a proper K-analytic map $h: Y \to X$ such that h gives a K-bianalytic map of $Y - h^{-1}(C_f)$ to $X - C_f$; there exists a finite set $\mathcal{E} = \{E\}$ of closed K-analytic submanifolds of Y of codimension 1 meeting transversally such that the divisors of $f \circ h$ and

$h^*(dx_1 \wedge \ldots \wedge dx_n)$ are $\Sigma N_E E$ and $\Sigma(n_E - 1)E$ with N_E and n_E positive for every E in $\mathcal{E}$. We shall denote by $N(\mathcal{E})$ the nerve of $\mathcal{E}$; it is a simplicial complex such that a p-simplex corresponds to a subset of $\mathcal{E}$ of cardinality $p + 1$ with its members having a nonempty intersection. We observe that $N(\mathcal{E})$ is equipped with the function $E \to (N_E, n_E)$, called the numerical data of E, on the set of its vertices. For instance if $n = 2$ and $f(x) = x_1^3 + x_2^2$, then $N(\mathcal{E})$ for its minimal desingularization looks like the Dynkin diagram of type D_4; the numerical data of the three end vertices and the fourth vertex are (1,1), (2,2), (3,3), and (6,5).

We take ω from $\Omega(K^x)$ and put

$$\mathcal{E}(\omega) = \{E \in \mathcal{E};\ \Gamma(\omega^{N_E}\omega_{n_E}) = \infty\}, \qquad m_\omega = \dim . N(\mathcal{E}(\omega)) + 1.$$

If $K = \mathbb{R}, \mathbb{C}$, we define s as above, i.e., as $\omega(i) = |i|_K^s (i/|i|)^p$; if K is a p-adic field, we define s as $\omega(\pi) = \omega_s(\pi)$. We observe that if $\mathrm{card}(O_K/\pi O_K) = q$, then $t = q^{-s}$, not s itself, is uniquely defined. At any rate for a particular ω, say ω', we denote the corresponding s by s'. If now s is close to s', we have

$$Z_\Phi(\omega) = \sum_{j=1}^{m_{\omega'}} R_{\omega',j}(\Phi)(s - s')^{-j} + \text{holo. fn at } s = s'.$$

In the case where K is a p-adic field Langlands [23] has expressed each $R_{\omega',j}(\Phi)$ by a principal-value integral; if we denote $R_{\omega',j}(\Phi)$ for $j = m_{\omega'}$ by $R_{\omega'}(\Phi)$, Langlands' formula for $R_{\omega'}(\Phi)$ is roughly as follows:

We choose $m_{\omega'}$ distinct members of $\mathcal{E}(\omega')$ and take their intersection and then we take the union D of all such intersections; and finally we denote by D_o the difference of D and the union of all E not in $\mathcal{E}(\omega')$. Then out of the resolution data we can define a complex measure $\mu_{\omega'}$ on D_o such that

$$R_{\omega'}(\Phi) = PV\int_{D_o} (\Phi \circ h)\mu_{\omega'} .$$

The complex measure $\mu_{\omega'}$ involves special values of complex parameters; "PV" means that we start from a convergent integral depending holomorphically on parameters and take its holomorphic continuation. We might mention that a formula similar to the above also exists in the case where $K = \mathbb{R}, \mathbb{C}$ and it can be proved in the same way.

As a simple application we shall outline a transparent proof of the fact that

$$\sigma(f) = \min_{E \in \mathcal{E}} \{n_E/N_E\}$$

is intrinsic, i.e., independent of the choice of $h: Y \to X$: if $\omega' = \omega_{-\sigma(f)}$, then D_o is not empty, $\mu_{\omega'}$ becomes a positive measure on D_o with support D_o, and the integral is absolutely convergent, hence "PV" can be dropped; this is so including the archimedean case. Therefore $R_{\omega'}(\Phi) > 0$ for some $\Phi > 0$. In other words $Z_\Phi(\omega)$ is holomorphic on $\Omega_{-\sigma(f)}(K^\times)$ and has a pole at $\omega' = \omega_{-\sigma(f)}$ for a suitable Φ. Therefore $\sigma(f)$ is intrinsic. It is an interesting problem in algebraic geometry to extract further intrinsic objects out of the complex $N(\mathcal{E})$.

§5. Theorems of Bernshtein and Denef

In the archimedean case Bernshtein [5] has found another way to prove the meromorphic continuation of $Z_\Phi(\omega)$; we shall first explain his theorem in precise terms:

"Let F_o denote any field of characteristic 0, $F = F_o(s)$ its simple transcendental extension, and $F[x, \partial/\partial x]$ the ring of linear differential operators

$$\Sigma\, a_{i_1 \ldots j_n} x_1^{i_1} \ldots x_n^{i_n} \partial^{j_1 + \ldots + j_n} / \partial x_1^{j_1} \ldots \partial x_n^{j_n}$$

with $a_{i_1 \ldots j_n}$ in F; take any $f \neq 0$ from $F_o[x] = F_o[x_1, \ldots, x_n]$ and convert the ring of fractions $F[x]_f = F[x, 1/f]$ of $F[x]$ into an $F[x, \partial/\partial x]$-module as

$$x_i \cdot \phi = x_i \phi, \quad (\partial/\partial x_i) \cdot \phi = \partial\phi/\partial x_i + (s\phi) f^{-1} \partial f/\partial x_i \quad (1 \le i \le n).$$

Then there exists an element P of $F[x, \partial/\partial x]$ satisfying $P \cdot f = 1$."

If we consider the set of all $B(s)$ in $F_o[s]$ of the form $P \cdot f$ for some P in $F_o[s, x, \partial/\partial x]$, we get an ideal different from zero; the unique monic generator $b_f(s)$ of this ideal is called the

Bernshtein polynomial of f. For instance in the example $f(x) = x_1^3 + x_2^2$ already mentioned in §4, if we put

$$P = (1/27)\partial^3/\partial x_1^3 + (1/6)x_1\partial^3/\partial x_1\partial x_2^2 + (1/8)x_2\partial^3/\partial x_2^3 + (3/8)\partial^2/\partial x_2^2,$$

we obviously have

$$P\cdot f = (s + 1)(s + 5/6)(s + 7/6);$$

and, e.g., by some results in §8 we can show that this is $b_f(s)$.

We take $\mathbb{R}$ as F_0 and denote by D a nonempty subset of $\mathbb{R}^n$ with the property that $f(x) > 0$ on D and $f(x) = 0$ on the boundary of D; then

$$f_D^s(\Phi) = \int_D f(x)^s\ \Phi(x)\ |dx|_{\mathbb{R}},$$

where Φ is in $S(\mathbb{R}^n)$, defines an $S(\mathbb{R}^n)'$-valued holomorphic function f_D^s on the right-half plane $\mathrm{Re}(s) > 0$. On the other hand, if x is in D, then $P\cdot f = b_f(s)$ can be rewritten as $Pf(x)^{s+1} = b_f(s)f(x)^s$. Therefore if P^* denotes the adjoint of P, by an induction on m we get

$$f_D^s(\Phi) = \prod_{i=0}^{m-1} (1/b_f(s + i))\cdot f_D^{s+m}((P^*)^m\Phi);$$

and consequently if we write

$$b_f(s) = \prod_\lambda (s + \lambda),$$

then $f_D^s/\Pi\Gamma(s + \lambda)$ becomes an $S(\mathbb{R}^n)'$-valued holomorphic function on the whole s-plane.

A p-adic counterpart of the above theorem has been and still is a problem. A recent result of Denef [9] is an important step in this direction; its exact statement is as follows:

"We define a family of three types of subsets of $\mathbb{Q}_p^n$ by using elements $f, g \neq 0$ of $\mathbb{Q}_p[x_1, \ldots, x_n]$: (i) $f(x) = 0$; (ii) $|f(x)|_p \leq |g(x)|_p$, where $|\ |_p = |\ |_{\mathbb{Q}_p}$; (iii) the projection to the first factor of $\mathbb{Q}_p^n \times \mathbb{Q}_p$ of its subset defined by $f(x) = y^m$ where $m \geq 2$. If D is a subset of $\mathbb{Q}_p^n$ with compact closure obtained by taking intersections, unions, and complements of the members of the above family a finite number of times, then

$$\int_D |f(x)|_p^s \, |dx|_p,$$

where $\mathrm{Re}(s) > 0$, represents a rational function of p^{-s}."

Denef has proved the above theorem in two ways, one using and another without using Hironaka's theorem, but using Macintyre's theorem [26]. According to what Professor Macintyre told us at the time of the conference, a generalization of his theorem to cover the case of positive characteristic is an open problem. At any rate, as an application, Denef has settled the following conjecture by Serre [43], p. 146:

"Let $f_1(x), \ldots, f_r(x)$ denote elements of $\mathbb{Z}_p[x_1, \ldots, x_n]$; then

$$\sum_{e=0}^{\infty} \mathrm{card}\{x \bmod p^e;\ x \in \mathbb{Z}_p^n,\ f_1(x) = \ldots = f_r(x) = 0\}\ t^e$$

represents a rational function of t."

We might mention that the rationality of a similarly defined series, where $f_1(x) = \ldots = f_r(x) = 0$ is replaced by $f_1(x) \equiv \ldots \equiv f_r(x) \equiv 0 \bmod p^e$, has been proved by Meuser [30].

In the case where $r = 1$, if $f(x)$ is in $O_K[x_1, \ldots, x_n]$ and

$$Z(t) = \int_{O_K^n} |f(x)|_K^s \, |dx|_K,$$

where $t = q^{-s}$ and $\mathrm{Re}(s) > 0$, we have

$$\sum_{e=0}^{\infty} \mathrm{card}\{x \bmod \pi^e; x \in O_K^n,\ f(x) \equiv 0 \bmod \pi^e\}(q^{-n}t)^e = (1-tZ(t))/(1-t).$$

Since $Z(t) = Z_\Phi(\omega_s)$, where Φ is the characteristic function of O_K^n, it is a rational function of t; cf. §4. In [19] we have computed $Z(t)$ in various cases and found the following experimental, hence conjectural, theorem:

"Suppose that $f(x) \neq 0$ in $O_K[x_1, \ldots, x_n]$ is homogeneous and has a good reduction $\bmod \pi$; then $\deg(Z(t)) + \deg(f(x)) = 0$."

For instance if $f(x)$ is a quartic invariant with coefficients in a number field k of the 56-dimensional irreducible representation of a connected simply connected simple group of type E_7, then for almost all $K = k_v$ we have

$$Z(t) = N(t)/(1 - q^{-1}t)(1 - q^{-11}t^2)(1 - q^{-19}t^2)(1 - q^{-28}t^2),$$

in which

$$N(t) = (1 - q^{-1})(1 - q^{-14})\{(1 + q^{-14}) - q^{-11}(1 + q^{-4} + q^{-8} - q^{-18})t + q^{-15}(1 - q^{-10} - q^{-14} - q^{-18})t^2 + q^{-30}(1 + q^{-14})t^3\} ;$$

hence $\deg(Z(t)) = 3 - 7 = -4 = -\deg(f(x))$.

§6. Prehomogeneous vector spaces

We shall go back to the case of group invariants: we take an affine space X, a connected reductive algebraic subgroup G of $GL(X)$, and an irreducible hypersurface H in X, all defined over a local field K, satisfying the basic assumption that $Y = X - H$ is a G-orbit. We choose a generator f, defined over K, of the ideal for H; then f is homogeneous and $f(gx) = \nu(g)f(x)$ for every g in G with a rational character ν of G defined over K. If we denote by G^1 the kernel of ν, we have $G/G^1 \cong GL_1$. We shall assume that G^1 has only a finite number of orbits in H; then each G^1-orbit in H is a G-orbit. We shall finally assume that there exists an involution $a \to a^\iota$ of $End(X)$ defined over K under which G is stable; then there exists a symmetric or alternating bilinear form $[x,y]$ on X defined over K such that $[x,ay] = [a^\iota x,y]$ for every a in $End(X)$. By an early theorem of Mostow [32] this is not an assumption if $K = \mathbb{R}, \mathbb{C}$. In fact if $K = \mathbb{C}$, we may assume that G, X are defined over $\mathbb{Q}$, hence over $\mathbb{R}$; and if $K = \mathbb{R}$, there exists a positive-definite $\mathbb{R}$-bilinear form $[x,y]_{\mathbb{R}}$ on $X_{\mathbb{R}}$ under which $G_{\mathbb{R}}$ is self-adjoint. Since $G_{\mathbb{R}}$ is Zariski dense in G by Rosenlicht [36], we have only to take the bilinear extension of $[x,y]_{\mathbb{R}}$ to X as $[x,y]$.

The above setup is a minor variant of Sato's "prehomogeneous vector spaces"; cf., among others, [44], [39], [40]. We have

$$\det(g)^2 = \nu(g)^{2\varkappa}, \quad \varkappa = \dim(X)/\deg(f)$$

for every g in G, in which $2\varkappa$ is an integer.

We take a Haar measure dx on X_K and define a Fourier transformation $\Phi \to \Phi^*$ in $S(X_K)$ as

$$\Phi^*(x) = \int_{X_K} \psi_K([x,y])\Phi(y)dy.$$

We shall normalize dx so that we get $(\Phi^*)^*(x) = \Phi(\pm x)$ according as $[x,y]$ is alternating or symmetric; and we introduce a G_K-invariant measure μ on Y_K as

$$\mu(x) = |f(x)|_K^{-\varkappa}\, dx.$$

Also for any T in $S(X_K)'$ we define T^* as in §1, i.e., as $T^*(\Phi) = T(\Phi^*)$. We know that Y_K splits into a finite number of G_K-orbits, say $Y_1, \ldots, Y_\ell$; cf. Serre [41], III-33. We take ω from $\Omega_\varkappa(K^\times)$ and put

$$Z_i(\omega)(\Phi) = \int_{Y_i} \omega(f(x))\Phi(x)\mu(x)$$

for $1 \leq i \leq \ell$; then $Z_i(\omega)$ becomes an $S(X_K)'$-valued holomorphic function on $\Omega_\varkappa(K^\times)$, and we have the following theorem:

"Each $Z_i(\omega)$ has a meromorphic continuation to the whole $\Omega(K^\times)$ and satisfies

$$Z_i(\omega)^* = \sum_{j=1}^{\ell} \gamma_{ij}(\omega) Z_j(\omega_\varkappa \omega^{-1})$$

with ℓ^2 meromorphic functions $\gamma_{ij}(\omega)$ on $\Omega(K^\times)$. Furthermore, if K is a p-adic field, $Z_i(\omega)(\Phi)$ and $\gamma_{ij}(\omega)$ are rational functions of $t = \omega(\pi)$ for each $\omega|U_K$."

We observe that $Z_1(\omega), \ldots, Z_\ell(\omega)$ and the system of functional equations for them generalize (ZL) and (FL) in §1. If $K = \mathbb{R}$, the above theorem in a slightly different formulation is in [44], [39] both on p. 142; if K is a p-adic field, it is in [19]. Sato's proof in the case where $K = \mathbb{R}$ depends on the observation that a differential operator "P" satisfying "$Pf(x)^{s+1} = b(s)f(x)^s$," cf. §5, can be found explicitly in this case: as we have recalled, we can find a coordinatization $\theta: X \cong \mathrm{Aff}^n$ defined over K such that G becomes selfadjoint with respect to $(x,y) \to {}^t\theta(x)\theta(y)$. If we put

$f(x) = f_\theta(\xi)$, where $\xi = \theta(x)$, then

$$f_\theta(\partial/\partial\xi) f_\theta(\xi)^{s+1} = ab(s) f_\theta(\xi)^s$$

for every s in $\mathbb{N}$ with a in $K^\times$ and with a monic polynomial $b(s)$ in s of degree equal to $\deg(f)$. Furthermore $b(s)$ depends only on G, and in particular it has rational coefficients. The intrinsic polynomial $b(s)$ so defined is called the Sato polynomial; by definition $b_f(s)$ divides $b(s)$. A list of $b(s)$ in the case where G is irreducible can be found in Kimura [22] and Ozeki [35].

If we put

$$b(s) = \prod_\lambda (s + \lambda), \quad \gamma(s) = \prod_\lambda \Gamma(s + \lambda - \kappa),$$

then in the case where $K = R$ and $\omega(i) = |i|_K^s (i/|i|)^p$ we see as in §5 that $Z_i(\omega)/\gamma(s)$ and $\gamma_{ij}(\omega)/\gamma(s)$ become holomorphic functions on the whole s-plane. Still in the case where $K = \mathbb{R}$ Sato's theory provides almost complete information on $\gamma_{ij}(\omega)$; cf. loc. cit. In the p-adic case we only know the poles of $Z_i(\omega)$ and $\gamma_{ij}(\omega)$; cf. op. cit. A progress will be made in this case if we can prove the "experimental theorem" in §5.

We might mention that $\gamma_{ij}(\omega_s)$ have been computed in several cases; cf. [44], [19], [7]. If $\ell > 1$, the expressions for $\gamma_{ij}(\omega_s)$ have turned out to be rather complicated. However in all known cases we have the following simple experimental theorem:

"Suppose that $K = \mathbb{C}$ or, more generally, that $\ell = 1$ and G splits over K; then $\gamma_{11}(\omega_s) = \prod_\lambda \Gamma(\omega_{s+\lambda-\kappa})$, in which $b(s) = \prod_\lambda (s + \lambda)$."

We might recall that, according to the definition in §1, we have

$$\Gamma(\omega_s) = \begin{cases} (2d\pi)^{d(1-2s)} \Gamma(ds)/\Gamma(d(1-s)) & d = \frac{1}{2}[K:\mathbb{R}] \\ (1 - q^{-(1-s)})/(1 - q^{-s}) & K \text{ p-adic field.} \end{cases}$$

§7. Zeta functions

If we have the same situation as in §6 where K is replaced by a number field k, we can introduce the following zeta function:

$$Z(\omega)(\Phi) = \int_{G_A/G_k} (\sum_{\xi\in Y_k} \Phi(g\xi))\omega(\nu(g))\mu(g),$$

in which ω is in $\Omega_\sigma(I_k/k^\times)$ for a large σ, Φ is in $S(X_A)$, and μ is a Haar measure on G_A; this is a generalization of (ZG2) in §1. We observe that if we define Φ^g, gT for any bicontinuous automorphism g of X_A and for any Φ, T in $S(X_A)$, $S(X_A)'$ as $\Phi^g(x) = \Phi(gx)$, $(gT)(\Phi) = T(\Phi^g)$, we have

$$gZ(\omega) = \omega(\nu(g))^{-1}Z(\omega)$$

for every g in G_A. Furthermore if $f(x)$ is an invariant in the classification of §2 where "r" = 1, very roughly $Z(\omega)(\Phi)$ is the Mellin transform of $I'(\Phi^t)$ for t in I_k. At any rate if G is the identity component of the group of all similarities of $f(x)$, such a zeta function has been examined by Weil [48] in the case of a nondegenerate quadratic form, by Mars [28] in the case of a cubic form mentioned in §2, and by us in other cases. In all cases $Z(\omega)$ has a meromorphic continuation to the whole $\Omega(I_k/k^\times)$ and satisfies the functional equation

$$Z(\omega)^* = Z(\omega_\varkappa\, \omega^{-1}),$$

which is a generalization of (FG) in §1. And, e.g., if $f(x)$ is the quartic invariant in §5, called the (original) Freudenthal quartic, we have the following additional information:

We recall that there exists an alternating bilinear form $[x,y]$ on $X = \mathrm{Aff}^{56}$ such that $[gx,gy] = \nu_o(g)[x,y]$ for every g in G, in which ν_o is a rational character of G with ν as its square. If we denote by G_o^1 the kernel of ν_o, then G becomes a semidirect product of G_o^1 by GL_1; and G_o^1 is a connected simply connected simple group of type E_7. We may assume that μ is the product of the Tamagawa measure on $(G_o^1)_A$ and the normalized Haar measure on I_k in §1. If we put

$$Z_o(\omega)(\Phi) = \int_{G_A/G_k} (\sum_{\xi\in Y_k} \Phi(g\xi))\omega(\nu_o(g))\mu(g),$$

we obviously have $Z(\omega) = Z_o(\omega^2)$; and we can show that $Z_o(\omega)^* = Z_o(\omega_{28}\, \omega^{-1})$. Furthermore if $\omega|I_k^1 \neq 1$, then $Z_o(\omega)$ is holomorphic on the whole $\Omega(I_k/k^\times)$ while if $\omega|I_k^1 = 1$, i.e., if

$\omega = \omega_s$, it has poles of order 1 at 0, 2, 9, 11, 17, 19, 26, 28. In other words $Z(\omega_s)$ has poles of order 1 at 0, 1, $4\frac{1}{2}$, $5\frac{1}{2}$, $8\frac{1}{2}$, $9\frac{1}{2}$, 13, 14. We observe that if $b(s) = \Pi(s + \lambda)$, then they are the λ's and $(\varkappa - \lambda)$'s; cf. Kimura [22], p. 78. At any rate the residues of $Z(\omega_s)$ at $s = 13, 14$ are respectively $\frac{1}{2}$-times

$$- \int_{U(0)_A} \Phi(x)\,|\theta_0(x)|_A, \quad \tau_k(G_o^1)\int_{X_A} \Phi(x)\,|dx|_A .$$

We might mention that in general the functional equation for $Z(\omega)$ is easier to obtain than the Siegel-Weil formula because neither precise information on Tamagawa numbers nor the Poisson formula $(P^{\#})$ in §2 is needed; the classical Poisson formula (P) in §1 is enough. We recall that the Siegel-Weil formula is open for a Freudenthal quartic.

A nonadelic theory of similar zeta functions in the more general prehomogeneous case is in Sato-Shintani [39] and in its special case where $f(x)$ is the discriminant of a binary cubic form is in Shintani [44] both for $k = \mathbb{Q}$. An adelic treatment of Shintani's case has been given by Wright [51] for any number field k and by Datskovsky [7] for a function field k, the case which we have excluded. The main interest comes from its connection with the arithmetic of binary cubic forms. We also mention that F. Sato [38] has examined zeta functions in the case where "H" is not necessarily irreducible.

On the other hand a generalization of (ZG1) in §1 has been given by Ono [34]: we take a polynomial $f(x)$ of degree m in n variables with coefficients in k, put $X = \mathrm{Aff}^n$, and denote by Y the complement of the hypersurface $f(x) = 0$, which we assume to be irreducible. Then the restricted-product measure of $|f(x)|_K^{-\varkappa}|dx|_K$, where $\varkappa = n/m$ and each multiplied by $(1 - q^{-1})^{-1}$ if $K = k_v$ is a p-adic field, gives a positive measure on Y_A. We further multiply a positive constant independent of $f(x)$ so that in the case where $f(x) = x$ we get the normalized Haar measure on I_k in §1. We shall denote the so-normalized positive measure on Y_A by μ' and we put

$$Z'(\omega)(\Phi) = \int_{Y_A} \omega(f(x))\Phi(x)\mu'(x).$$

Then $Z'(\omega)(\Phi)$ becomes a holomorphic function on $\Omega_\varkappa(I_k/k^\times)$; and it is continuous for $\sigma(\omega) \geq \varkappa$ if $\omega|I_k^1 \neq 1$ while

$$\lim_{s\to\varkappa} (s - \varkappa)Z'(\omega_s)(\Phi) = \int_{X_A} \Phi(x)\,|dx|_A$$

if $\omega|I_k^1 = 1$, hence $\omega = \omega_s$. Furthermore in the prehomogeneous case we have $gZ'(\omega) = \omega(\nu(g))^{-1}Z'(\omega)$ for every g in G_A and in some cases, such as Mars' case, $Z(\omega)$ and $Z'(\omega)$ differ by a constant factor.

If $f(x)$ is a Freudenthal quartic, we can show that $Z'(\omega_s)(\Phi)$ has a meromorphic continuation at least to $\mathrm{Re}(s) > 7$ and that it has the same residue as $2Z(\omega_s)(\Phi)$ at $s = 13$. If therefore $Z'(\omega_s) = 2Z(\omega_s)$, we will have $\tau_k(G_o^1) = 1$. However it seems too much to expect that the product of all $(1 - q^{-1})^{-1}N(t)$ in §5, which is an irreducible polynomial in $\mathbb{C}(q)[t]$ if q is considered as a variable, has a meromorphic continuation to the whole s-plane.

§8. Two variable local case

We shall finally explain a small success story about the following problem: we know that poles of $Z_\Phi(\omega)$ can be described by the resolution data; in fact we have Langlands' formula for the coefficients of the principal parts of its Laurent expansions. However since the formula involves "PV", it is not immediately clear whether or not a certain principal part simply represents 0, i.e., whether or not a certain pole is fictitious. The simplest case of some theoretical interest is the case where $f(x)$ is a polynomial in just two variables x_1, x_2 satisfying $f(0) = 0$ with 0 as its critical point and such that it is irreducible in $\overline{K}[[x_1,x_2]]$, in which $\overline{K}$ is an algebraic closure of K.

We shall be interested in the algebroid curve C around 0 defined by $f(x) = 0$ and accordingly we shall assume that $\mathrm{Supp}(\Phi)$ is contained in a small open neighborhood of 0 in K^2. Then there exists a well-known minimal desingularization of C, which is the product of a unique sequence of quadratic transformations. We shall denote the exceptional curves in the order of their creation by E_1, E_2, ... and include the strict transform of C as the last "E" so that we can write $\mathcal{E} = \{E_I\}_I$. Then $N(\mathcal{E})$ is a "tree", cf. Serre [42], with $g \geq 1$ branching vertices where three 1-simplices meet and with $g + 2$ end vertices; we recall that $N(\mathcal{E})$ is equipped with the function $E \to (N_E,n_E)$ or rather $I \to (N_I,n_I)$ on the set of its vertices. This function can be described by the characteristic pairs $(\mu_1,\nu_1), \ldots, (\mu_g,\nu_g)$ of C, which depend only on the factor ring of $\overline{K}[[x_1,x_2]]$ by the principal ideal generated by $f(x)$; cf. [16]. We at least recall that μ_i, ν_i are relatively prime integers satisfying $\nu_i \geq 2$ for all i, $\mu_1/\nu_1 > 1$, and $\mu_i/\nu_i - \mu_{i-1} > 0$ for $i > 1$ and that $\sigma(f)$, defined locally, is given by

$$\sigma(f) = (\mu_1 + \nu_1)/\mu_1\nu_1 \cdots \nu_g.$$

Now if E_I is a nonbranching vertex and if $E_{I'}$, $E_{I''}$ are its neighboring vertices with $E_{I''}$ created after E_I by $\rho - 1$ quadratic transformations, then

$$(N_{I'},n_{I'}) + (N_{I''},n_{I''}) = \rho(N_I,n_I).$$

On the other hand if E_I is a branching vertex and if $E_{I'}$, $E_{I''}$, $E_{I'''}$ are its neighboring vertices with $E_{I'''}$ created after E_I by $\rho - 1$ quadratic transformations, then

$$(N_{I'},n_{I'}) + (N_{I''},n_{I''}) + (N_{I'''},n_{I'''}) = \rho(N_I,n_I) + (0,1).$$

Furthermore, at least in the p-adic case, the first relation is responsible for the fact that E_I has no contribution to the poles of $Z_\Phi(\omega)$ and the second relation is responsible for the fact that E_I does have a contribution to the poles of $Z_\Phi(\omega_s)$.

These are the major results of Strauss [45] and Meuser [31] in a slightly generalized form. When we explained those in detail at Collège de France in May of 1983, Professor Serre mentioned that the taking of the sum of a function over neighboring vertices was a known operation in the theory of trees as Hecke operator, Lapalacian, etc. We recall that a harmonic function can be characterized by the property that its average over the surface of any small sphere is equal to its value at the center. Therefore, in a rather peculiar sense, the above relations may be regarded as the harmonicity and the nonharmonicity of the function $I \to (N_I,n_I)$ respectively at nonbranching and branching vertices.

We go back to $Z_\Phi(\omega_s)$ and state the results more precisely: firstly and above all poles are of order 1. We denote the branching vertices in the order of their creation by $E_{I_1}, \ldots, E_{I_g}$. In the p-adic case if we put $t = q^{-s}$, there are ΣN_{I_i} poles $\{\alpha\}$ and one more pole q in the t-plane, where

$$\alpha^{N_{I_i}} = q^{n_{I_i}} \quad (1 \le i \le g).$$

Furthermore

$$\lim_{t \to \alpha} (1 - \alpha^{-1}t) Z_{\Phi}(\omega_s) = m_K/N_{I_i} \cdot c_{\alpha}(f) \cdot \Gamma(\omega_{s'})\Gamma(\omega_{s''})\Gamma(\omega_{s'''}) \cdot \Phi(0),$$

in which $m_K = 1 - q^{-1}$, $c_{\alpha}(f) \neq 0$ depends on the coefficients of $f(x)$, and, e.g., $s' = n_{I'} - \lambda N_{I'}$, where $q^{\lambda} = \alpha$. In the case where $K = \mathbb{R}, \mathbb{C}$ if $\lambda = n_{I_i}/N_{I_i}$, $\dim . N(E(\omega_{-\lambda})) = 0$, and if no E_{I_j} is in $E(\omega_{-\lambda})$ for $j < i$, then

$$\lim_{s \to -\lambda} (s + \lambda) Z_{\Phi}(\omega_s) = \text{the same as above}$$

with $m_K = 2, \; 2\pi$ according as $K = \mathbb{R}, \mathbb{C}$. The product of the Γ's is different from 0 in both cases. These are in our preprint, "Complex powers of irreducible algebroid curves"; we have used Langlands' formula and a formula on the convolution of ω's in Sally-Taibleson [37]. In the case where $K = \mathbb{R}, \mathbb{C}$ the conditions are satisfied if N_{I_i} and n_{I_i} are relatively prime, hence the above universal residue formula in that case; this in an equivalent form for $K = \mathbb{R}$ is the major result of Lichtin [25].

In the p-adic case $Z_{\Phi}(\omega_s)$ is periodic with period $2\pi i/\log q$, hence q^{-s} is a natural parameter; and we know exactly where the poles of $Z_{\Phi}(\omega_s)$ are in the parameter plane. In the archimedean case the poles of $Z_{\Phi}(\omega_s)$ are distributed in a finite number of arithmetic progressions of negative rational numbers mod 1, hence $\mathfrak{e}(-s)$ is an appropriate parameter. The fact is that the product of all $t - \mathfrak{e}(\lambda)$ where $-\lambda$ is taken from each arithmetic progression of poles can be determined in the case where $K = \mathbb{C}$; the exact statement is as follows:

"We put $m_i = \nu_{i+1} \cdots \nu_g$ and

$$\Delta_f(t) = \prod_{i=1}^{g} P_{N_{I_i}/m_{i-1}, \nu_i}(t^{m_i}),$$

in which $P_{a,b}(t) = (t-1)(t^{ab}-1)/(t^a-1)(t^b-1)$; further we put

$$P_f(t) = \Delta_f(t)/\text{G.C.D.}\ \{\Delta_f(t),\ d\Delta_f(t)/dt\}.$$

Then the product of all $t - e(\lambda)$ explained above is $(t-1)P_f(t)$ if $K = \mathbb{C}$ and a factor of $(t-1)P_f(t)$ is $K = \mathbb{R}$."

Actually there is a more general theorem: let $f(x)$ denote a polynomial in $n \geq 2$ variables with coefficients in $\mathbb{C}$ satisfying $f(0) = 0$ with 0 as its isolated critical point; let $\Delta_f(t)$ (resp. $P_f(t)$) denote the characteristic (resp. minimal) polynomial of the local Picard-Lefschetz monodromy of f at 0; assume that $\Delta_f(1) \neq 0$ and write

$$P_f(t) = \prod_{0<\lambda<1} (t - e(\lambda));$$

then

$$Z_\Phi(\omega_s)/\Gamma(s+1)\prod_\lambda \Gamma(s+\lambda)$$

is holomorphic on the whole s-plane and no Γ-factor can be deleted without creating a pole. The statement under quotation follows from this theorem and a theorem of Lê Dũng Tráng [24] stating that the general $\Delta_f(t)$, $P_f(t)$ above become the explicit $\Delta_f(t)$, $P_f(t)$ in the quotation if $n = 2$. A proof of the above theorem is in our preprint, "Complex power of a hypersurface with isolated singularity." Finally Barlet [3] has a more general theorem, without any assumption on $f(x)$, stating that the product of $\Gamma(s+\lambda)$ defined by the minimal polynomial of the whole local Picard-Lefschetz monodromy of f over 0, not just its $(n-1)$-dimensional part as in our theorem, is at least necessary to kill all poles of $Z_\Phi(\omega_s)$; we refer to his paper for the details.

References

[1] H. Ariturk, The Siegel-Weil formula for orthogonal groups, Thesis, Johns Hopkins, 1975.

[2] M.F. Atiyah, Resolution of singularities and division of distributions, Comm. pure and appl. math. 23 (1970), 145-150.

[3] D. Barlet, *Contribution effective de la monodromie aux développements asymptotiques*, preprint.

[4] I.N. Bernshtein and S.I. Gel'fand, *Meromorphic property of the functions* P^λ, Functional Analysis and its Applications 3 (1969), 68-69.

[5] I.N. Bernshtein, *The analytic continuation of generalized functions with respect to a parameter*, Functional Analysis and its Applications 6 (1972), 273-285.

[6] A. Borel, *Some finiteness properties of adele groups over number fields*, Pub. math. I.H.E.S. 16 (1963), 5-30.

[7] B.A. Datskovsky, *On zeta functions associated with the space of binary cubic forms with coefficients in a function field*, Thesis, Harvard, 1984.

[8] P. Deligne, *La conjecture de Weil*. I, Pub. math. I.H.E.S. 43 (1974), 273-307.

[9] J. Denef, *The rationality of the Poincaré series associated to the p-adic points on a variety*, preprint.

[10] S.J. Haris , *An equality of distributions associated to families of theta series*, Nagoya Math. J. 59 (1975), 153-168.

[11] D.R. Heath-Brown, *Cubic forms in ten variables*, London Math. Soc. 47 (1983), 225-257.

[12] H. Hironaka, *Resolution of singularities of an algebraic variety over a field of characteristic zero*. I-II, Ann. Math. 79 (1964), 109-326.

[13] J. Igusa, *On certain representations of semi-simple algebraic groups and the arithmetic of the corresponding invariants*. I, Invent. math. 12 (1971), 62-94.

[14] J. Igusa, *Geometry of absolutely admissible representations*, Number Theory, Algebraic Geometry and Commutative Algebra, Kinokuniya, Tokyo (1973), 373-452.

[15] J. Igusa, *Complex powers and asymptotic expansions*, I. Crelles J. Math. 268/269 (1974), 110-130; II. ibid. 278-279 (1975), 307-321; or *Forms of Higher Degree*, Tata Inst. Lect. Notes 59, Springer-Verlag (1978).

[16] J. Igusa, *On the first terms of certain asymptotic expansions*, Complex Analysis and Algebraic Geometry, Iwanami Shoten, Tokyo (1977), 357-368.

[17] J. Igusa, *Exponential sums associated with a Freudenthal quartic*, J. Fac. Sci. Univ. Tokyo 24 (1977), 231-246.

[18] J. Igusa, *On Lie algebras generated by two differential operators*, Manifolds and Lie groups, Progress in Math. 14, Birkhäuser (1981), 187-195.

[19] J. Igusa, *Some results on p-adic complex powers*, to appear in the American Journal of Mathematics.

[20] P. Jeanquartier, *Développement asymptotique de la distribution de Dirac attachée à une fonction analytique*, C.R. 271 (1970), 1159-1161.

[21] V.G. Kac, *Infinite dimensional Lie algebras: an introduction*, Progress in Math. 44, Birkhäuser (1983).

[22] T. Kimura, The b-functions and holonomy diagram of irreducible regular prehomogeneous vector spaces, Nagoya Math. J. 85 (1982), 1-80.

[23] R.P. Langlands, Orbital integrals on forms of SL(3). I, Amer. J. Math. 105 (1983), 465-506.

[24] Lê Dũng Trang, Sur les noeuds algébriques, Comp. Math. 25 (1972), 281-321.

[25] B. Lichtin, Some algebro-geometric formulae for poles of $|f(x,y)|^s$, to appear in the American Journal of Mathematics.

[26] A. Macintyre, On definable subsets of p-adic fields, J. Symb. Logic, 41 (1976), 605-610.

[27] B. Malgrange, Intégrales asymptotiques et monodromie, Ann. Éc. Norm. Sup. 7 (1974), 405-430.

[28] J.G.M. Mars, Les nombres de Tamagawa de certains groupes exceptionnels, Bull. Soc. Math. France 94 (1966), 97-140.

[29] J.G.M. Mars, The Tamagawa number of ^2A_n, Ann. Math. 89 (1969), 557-574.

[30] D. Meuser, On the rationality of certain generating functions, Math. Ann. 256 (1981), 303-310.

[31] D. Meuser, On the poles of a local zeta function for curves, Invent. math. 73 (1983), 445-465.

[32] G.D. Mostow, Self-adjoint groups, Ann. Math. 62 (1955), 44-55.

[33] T. Ono, On the relative theory of Tamagawa numbers, Ann. Math. 82 (1965), 88-111.

[34] T. Ono, An integral attached to a hypersurface, Amer. J. Math. 90 (1968), 1224-1236.

[35] I. Ozeki, On the micro-local structure of the regular prehomogeneous vector space associated with $SL(5) \times GL(4)$. I, Proc. Japan Acad. 55 (1979), 37-40.

[36] M. Rosenlicht, Some rationality questions on algebraic groups, Annali di Mat. 43 (1957), 25-50.

[37] P.J. Sally and M.H. Taibleson, Special functions on locally compact fields, Acta Math. 116 (1966), 279-309.

[38] F. Sato, Zeta functions in several variables associated with prehomogeneous vector spaces. I, Tôhoku Math. J. 34 (1982), 437-483.

[39] M. Sato and T. Shintani, On zeta functions associated with prehomogeneous vector spaces, Ann. Math. 100 (1974), 131-170.

[40] M. Sato and T. Kimura, A classification of irreducible prehomogeneous vector spaces and their relative invariants, Nagoya Math. J. 65 (1977), 1-155.

[41] J.-P. Serre, Cohomologie Galoisienne, Lect. Notes in Math. 5, Springer-Verlag (1965).

[42] J.-P. Serre, Arbres, amalgames, SL_2, astérisque 46 (1977).

[43] J.-P. Serre, Quelques applications de théorème de densité de Chebotarev, Pub. math. I.H.E.S. 54 (1981), 123-201.

[44] T. Shintani, On Dirichlet series whose coefficients are class numbers of integral binary cubic forms, J. Math. Soc. Japan 24 (1972), 132-188.

[45] L. Strauss, Poles of a two-variable p-adic complex power, Trans. Amer. Math. Soc. 278 (1983), 481-493.

[46] T. Tamagawa, Adèles, Proc. Symp. pure Math. 9 (1966), 113-121.

[47] J. Tate, Fourier analysis in number fields and Hecke's zeta-functions, Thesis, Princeton, 1950; Algebraic Number Theory, Acad. Press (1967), 305-347.

[48] A. Weil, Adeles and algebraic groups, Institute for Advanced Study, 1961; Progress in Math. 23, Birkhäuser (1982).

[49] A. Weil, Sur la formule de Siegel dans la théorie des groupes classiques, Acta Math. 113 (1965), 1-87; Collected Papers III, Springer-Verlag (1979), 71-157.

[50] A. Weil, Fonction zêta et distributions, Sém. Bourbaki 312 (1966), 1-9; Collected Papers III, Springer-Verlag (1979), 158-163.

[51] D.J. Wright, Dirichlet series associated with the space of binary cubic forms with coefficients in a number field, Thesis, Harvard, 1982.

THE JOHNS HOPKINS UNIVERSITY
BALTIMORE, MARYLAND 21218

DEFORMATION SPACES ASSOCIATED TO COMPACT HYPERBOLIC MANIFOLDS

Dennis Johnson and John J. Millson*

In this paper we take a first step toward understanding representations of cocompact lattices in $SO(n,1)$ into arbitrary Lie groups by studying the deformations of rational representations - see Proposition 5.1 for a rather general existence result. This proposition has a number of algebraic applications. For example, we remark that such deformations show that the Margulis Super-Rigidity Theorem, see [30], cannot be extended to the rank 1 case. We remark also that if $\Gamma \subset SO(n,1)$ is one of the standard arithmetic examples described in Section 7 then Γ has a faithful representation ρ' in $SO(n+1)$, the Galois conjugate of the uniformization representation, and Proposition 5.1 may be used to deform the direct sum of ρ' and the trivial representation in $SO(n+2)$ thereby constructing non-trivial families of irreducible orthogonal representations of Γ. However, most of this paper is devoted to studying certain spaces of representations which are of interest in differential geometry in a sense which we now explain.

Recently, there has been considerable interest in spaces of locally homogeneous (or geometric) structures on smooth manifolds, see for example, Thurston [25]. The spaces of conformal, projective and hyperbolic structures are of particular interest. If M is a smooth manifold we will denote the corresponding spaces of <u>marked</u> structures, see Lok [13], page 7, by $\mathcal{C}(M)$, $\mathcal{P}(M)$ and $\mathcal{H}(M)$ respectively. Since these spaces are a measure of the complexity of the fundamental group, it makes sense to study them in the case that M is a hyperbolic n-manifold. Of course, if $n \geq 3$ and M is compact, then the celebrated Mostow Rigidity Theorem states that $\mathcal{H}(M)$ is a point. Our main theme is that this is far from true for $\mathcal{C}(M)$ and $\mathcal{P}(M)$. Also

*The first author was partially supported by NSF grant #MCS77-24103, the second by NSF grant #MCS-8200639.

$\mathcal{H}(M \times \mathbb{R})$ is an interesting space closely related to $\mathcal{C}(M)$. Our first main result is a lower bound for the dimensions of the three previous deformation spaces by r, the largest number of disjoint, non-singular, totally geodesic hypersurfaces contained in M. If M is a hyperbolic surface of genus g then $r = 3g - 3$. From this bound, it is easily shown that the deformation spaces have arbitrarily large dimension as M varies. Our second main result is the existence of non-isolated singularities. In fact, we prove that the deformation spaces are locally homeomorphic to certain singular algebraic varieties; however, it should be possible to prove that the deformation spaces themselves have natural local analytic structures (see the remark at the end of Section 7) preserved by the local homeomorphism hol (see below). We would then have established that $\mathcal{C}(M)$, $\mathcal{P}(M)$ and $\mathcal{H}(M \times \mathbb{R})$ are singular for their natural local analytic structures.

To obtain the above results concerning the spaces of structures it is convenient to replace them with the space of classes of representations of Γ, the fundamental group of M, into the automorphism groups of the model space. This is possible because of the following general result.

Let $\mathcal{S}(M)$ be a space of marked locally homogeneous structures modeled on a homogeneous space $X = G/H$. Given a structure $s \in \mathcal{S}(M)$, by continuing coordinate charts around elements of Γ, see Lok [13], page 6, we obtain the holonomy representation $\rho:\Gamma \to G$ of s and a map:

$$\text{hol}: \ \mathcal{S}(M) \to \text{Hom}(\Gamma,G)/G$$

defined by $\text{hol}(s) = G\cdot\rho$ where G acts by conjugation. Then Theorem 1.11 of Lok [13] states that hol is an open map which lifts to a local homeomorphism from the space of (G,X)-developments to $\text{Hom}(\Gamma,G)$. We will refer to this result as the "Holonomy Theorem". Unfortunately hol is not necessarily a local homeomorphism. However, if ρ is a stable representation (see Section 1) then there exist neighborhoods U of s in $\mathcal{S}(M)$ and V of ρ in $\text{Hom}(\Gamma,G)/G$, finite groups H_1 and H_2 with $H_1 \subset H_2$ (the isotropy subgroups of s and ρ) and finite quotient mappings $U = \tilde{U}/H_1$, $V = \tilde{V}/H_2$ such that hol lifts to a homeomorphism $\tilde{U}$ to $\tilde{V}$. In particular if ρ is good (see Section 1) then hol is a homeomorphism from a neighborhood of ρ to a neighborhood of ρ. We see then that if ρ is stable then local information around ρ gives us information around s.

The representation $\rho:\Gamma \to G$ is necessarily rigid because $\rho(\Gamma)$ is not necessarily a lattice in G. Thus, we circumvent the Mostow Rigidity Theorem by, on the one hand, considering second-order structures such as conformal and projective structures and on the other, by considering non-lattice subgroups. For the three deformations spaces considered above we have $G = SO(n+1,1)$ for $\mathcal{C}(M)$, $G = PGL_{n+1}(\mathbb{R})$ for $\mathcal{P}(M)$ and $G = SO(n+1,1)$ for $\mathcal{H}(M \times \mathbb{R})$. Thus, we can concentrate our efforts on the two families of spaces $Hom(\Gamma,SO(n+1,1))$ and $Hom(\Gamma,PGL_{n+1}(\mathbb{R}))$ and their quotients by $SO(n+1,1)$ and $PGL_{n+1}(\mathbb{R})$ respectively. With the exception of Section 8, this paper is entirely concerned with these latter spaces. In addition to proving the previous lower bound for the dimensions of these spaces, we give examples where they are singular at certain representations, including irreducible ones, for their natural algebraic structures.

It seems the first result showing the non-triviality of $Hom(\Gamma,SO(n+1,1))$ for $n \geq 3$ was Apanasov [1]. The matter was greatly clarified by Thurston's idea of bending a Fuchsian group, see Sullivan [24] or Kourouniotis [27].

There are a number of technical theorems contained in this paper in addition to the main results alluded to above. For the reader's convenience we briefly state them in order of occurrence.

Section 1 defines stable representations, characterizes them in terms of parabolic subgroups and proves they are Zariski open in $Hom(\Gamma,G)$. A slice theorem is proved for the action of G on the stable representations. A very general notion of quasi-Fuchsian representation is studied and found to be surprisingly restrictive.

Section 2 treats deformations and infinitesimal deformations of representations and the first obstruction to integrating an infinitesimal deformation. We study this obstruction via the dual homology class in later sections.

Section 3 deals with quasi-Fuchsian representations of $\Gamma = \pi_1(M)$ on hyperbolic $(n+1)$-space and our main theorem in this section shows they are open in $Hom(\Gamma,SO(n+1,1))$. We prove various theorems concerning the local nature of the space of conjugacy classes of quasi-Fuchsian representations; for example, for n even, this space is an open subset of the real algebraic set $X(\Gamma,SO(n+1,1))$ - see Section 1 for notation. In the odd case, this result is not necessarily correct, there is another component of the real points passing through the uniformization representation ρ of Γ in $SO(n,1)$ corresponding to deformations in the group $SO(n,2)$ of ρ composed with the inclusion of $SO(n,1)$ in $SO(n,2)$.

Section 4 is a technical section dealing with cycles with coefficients and their intersection products. This material is needed to compute the first obstruction class.

Section 5 is one of the main sections of the paper. We introduce an algebraic version of Thurston's bending deformation - see Kourouniotis [27] for a geometric definition justifying the name "bending". Theorem 5.1 identifies the derivative of the bending deformation with the Poincaré dual of a totally geodesic hypersurface with an obvious coefficient from Minkowski space. The rest of the section is concerned with proving that $\dim X(\Gamma, SO(n+1,1))$ and $\dim X(\Gamma, PGL_{n+1}(\mathbb{R}))$ are bounded below by r, the maximum number of disjoint, embedded, totally geodesic hypersurfaces in M. In the classical case of a hyperbolic surface of genus g we have $r = 3g - 3$ and the bound is a weak one. By a simple geometric construction the problem is reduced to deforming a representation in G of the fundamental group of a graph of groups such that all edge groups have non-zero invariants in $\mathfrak{g}$, the Lie algebra of G.

Section 6 gives a criterion for the above spaces to have non-isolated singularities. This criterion involves computing some intersections of cycles with coefficients. It is possible that the space $C(M)$ is singular for <u>any</u> hyperbolic n-manifold $(n \geq 4)$ admitting two different intersecting, two-sided, non-singular, totally geodesic hypersurfaces. However, we have not been able to prove this.

Section 7 is a technical section proving the existence of nicely intersecting totally geodesic submanifolds in the standard arithmetic examples. For example, we show (Theorem 7.2) that if $p + q \neq n - 1$ there exist totally geodesic submanifolds of codimension p and q respectively intersecting in a single component. The reader may find this section difficult - he is advised to refer to O'Meara [18] for background information on the Strong Approximation Theorem and the spinor norm.

Section 8 is concerned with the interaction of $C(M)$ with Riemann geometry. We state a theorem suggested to us by Jim Simons and proved by S.Y. Cheng which shows each conformal class of metrics on M contains a unique metric of constant <u>scalar</u> curvature $-n(n-1)$. (This also follows as a special case of the Yamabe problem, recently solved by R. Schoen [29].) This is a generalization of the General Uniformization Theorem for Riemann surfaces. Using this metric we construct an interesting function vol: $\mathcal{C}(M) \to \mathbb{R}_+$ which assigns to a conformal structure the volume of M for the canonical metric belonging to that structure. We prove that vol is not constant if $n \geq 3$ (it is constant if $n = 2$). In case $n = 4$, we prove that vol has an absolute minimum at the hyperbolic structure. For all $n \geq 3$, it has a local minimum (with positive definite Hessian) at the hyperbolic structure. The existence of the canonical metric combined with work of Gasqui and Goldschmidt [10] yields a Riemannian metric on $\mathcal{C}(M)$, generalizing the Petersson-Weil metric.

There are a great many problems concerning $\mathcal{C}(M)$ and $\mathcal{P}(M)$ which remain unanswered - their topological properties for example. It would be very useful to have some examples, for instance for some hyperbolic 3-manifolds. We believe that the most important problem is to decide whether r is always equal to the dimensions of $\mathcal{C}(M)$ and $\mathcal{P}(M)$ or just a lower bound. A closely related problem is to construct more deformations - perhaps by analytic methods.

We would like to thank a number of people who helped us with this paper. Above all, we thank Bill Goldman for suggesting the main lines of Theorem 3.1 and many other conversations. Also we would like to thank John Morgan for suggesting the proof of Lemma 3.4, and for suggesting the graph of hypersurfaces of Section 5, Larry Lok for providing us with his thesis and an extension of an argument of his thesis (see the proof of Theorem 3.1), Robert Steinberg for providing us with the proof of Lemma 1.1 and S.Y. Cheng for proving Theorem 8.1 and a helpful conversation concerning Theorem 8.3. We should acknowledge a debt to Bill Thurston, for his idea of bending a Fuchsian group is at the heart of this paper. Finally, this paper is dedicated to Dan Mostow on the occasion of his sixtiety birthday (the second author presented it at the conference at Yale marking this occasion). The second author would like to take this occasion to thank Dan Mostow and the Yale mathematics faculty for the hospitality shown him as a visitor in 1983-84, as an assistant professor from 1974-78 and on many other occasions.

After we had finished writing this paper we learned of the thesis of Kourouniotis [27]. Kourouniotis also obtains the lower bound for the dimension of $\mathcal{C}(M)$. His thesis contains a careful description of the geometric version of bending.

1. Character Varieties and Generalized Quasi-Fuchsian Groups.

Let Γ be a finitely generated group and $\underline{G}$ a simple linear algebraic group defined over $\mathbb{R}$. The complex points of $\underline{G}$ will also be denoted $\underline{G}$ and the real points G. The set $\mathrm{Hom}(\Gamma,\underline{G})$ is the set of complex points of an affine variety defined over $\mathbb{R}$ with real points $\mathrm{Hom}(\Gamma,G)$. We will often denote $\mathrm{Hom}(\Gamma,\underline{G})$ by $(\Gamma,\underline{G})$ and $\mathrm{Hom}(\Gamma,G)$ by (Γ,G). The group $\underline{G}$ acts algebraically on $(\Gamma,\underline{G})$ by conjugation. This action will be denoted $g\cdot\rho$ for $g \in \underline{G}$ and $\rho \in R(\Gamma,\underline{G})$. Since $\underline{G}$ is reductive, there is a quotient variety $X(\Gamma,\underline{G})$ for this action, see Newstead [19]. The set of real points of the quotient variety will be denoted $X(\Gamma,G)$. We let $\underline{\pi}:R(\Gamma,\underline{G}) \to X(\Gamma,\underline{G})$ and $\pi:R(\Gamma,G) \to X(\Gamma,G)$ denote the quotient projections.

The quotient variety $X(\Gamma,\underline{G})$ is obtained as follows. Suppose $\{\gamma_1,\gamma_2,\ldots,\gamma_N\}$ is a set of generators for Γ and $\{f_1,f_2,\ldots,f_m\}$ is a set of algebra generators for the algebra of invariant polynomials on $\underline{G}^N$. We may choose the f_i's so that they take real values on G^N. We define a map $F:R(\Gamma,\underline{G})/\underline{G} \to \mathbb{C}^m$ by:

$$F(\underline{G}\cdot\rho) = (f_1(\rho(\gamma_1),\ldots,\rho(\gamma_N)),\ldots,f_m(\rho(\gamma_1),\ldots,\rho(\gamma_N)))$$

We caution the reader that F is not necessarily injective.

The image of F is contained in an affine variety determined by the relations among the generators $\{\gamma_1,\ldots,\gamma_N\}$ and the relations among the invariants $\{f_1,\ldots,f_m\}$. Precisely, $X(\Gamma,\underline{G})$ is the affine variety corresponding to the ring of $\underline{G}$ invariant polynomials on $\mathrm{Hom}(\Gamma,\underline{G})$. Then $X(\Gamma,\underline{G})$ is defined over $\mathbb{R}$. The set of real points $X(\Gamma,G)$ contains the image under F of the classes of representations on which the invariants $\{f_1,f_2,\ldots,f_m\}$ take real values. We note that F is the mapping induced by $\underline{\pi}$ on the orbit space of $R(\Gamma,\underline{G})$ to $X(\Gamma,\underline{G})$.

As we have remarked previously, the variety $X(\Gamma,\underline{G})$ is not isomorphic to the orbit space $R(\Gamma,\underline{G})/\underline{G}$. However, we now define a subset of $R(\Gamma,\underline{G})$, the set $S = S(\Gamma)$ of stable representations. This set has the property that F induces a homeomorphism from $S/\underline{G}$ onto an open subset of $X(\Gamma,\underline{G})$.

Definitions. A representation ρ in $R(\Gamma,\underline{G})$ is said to be stable if the orbit $\underline{G}\cdot\rho$ is closed in $R(\Gamma,\underline{G})$ and if the isotropy subgroup $Z(\rho)$ of ρ in G is finite.

A stable representation is said to be good if $Z(\rho) = Z_{\underline{G}}$, the center of $\underline{G}$.

By Newstead [19], Proposition 3.8, S is Zariski open in $R(\Gamma,\underline{G})$. However, S might be empty.

Let Γ_N be the free group on N generators. Then we have a closed embedding $R(\Gamma,\underline{G}) \subset R(\Gamma_N,\underline{G})$. Our definition of stability then gives:

$$S(\Gamma) = R(\Gamma,\underline{G}) \cap S(\Gamma_N).$$

This equality allows us to reduce many problems for Γ to the corresponding problems for Γ_N. This is helpful because $R(\Gamma_N,\underline{G}) = \underline{G}^N$, a non-singular variety.

We now characterize the stable representations in terms of complex parabolic subgroups. Recall that a parabolic subgroup P of a semi-simple Lie group G is the full normalizer of a parabolic subalgebra - an algebra whose complexification contains a maximal solvable subalgebra - see Varadarajan [26], 279-288.

<u>Theorem 1.1</u>. <u>A representation</u> ρ <u>in</u> $R(\Gamma,G)$ <u>is stable if and only if the image of</u> ρ <u>is not contained in any proper parabolic subgroup of</u> G.

<u>Proof</u>. Assume ρ is stable. If the image of ρ is contained in a proper parabolic subgroup, then by conjugating ρ by a one parameter group in the center of a Levi subgroup we find a representation in the closure of the orbit of ρ which is contained in the Levi subgroup. Since the orbit of ρ is closed, this limit representation is conjugate to ρ. But the limit representation has an infinite centralizer in $\underline{G}$ (the centralizer contains the center of the Levi subgroup) contradicting the stability of ρ.

Now suppose that ρ is not stable. Hence, either the orbit of ρ is not closed or $Z(\rho)$ is not finite. Assume the former. By the Hilbert-Mumford Theorem, see Birkes [4], there is a one-parameter group $\lambda:\mathbb{C}_m \to \underline{\mathbb{C}}$ so that $\lim_{t\to 0}\lambda(t)\cdot\rho$ exists. By Mumford-Fogarty [17], Proposition 2.6, this implies that the image of ρ is contained in the parabolic group $P(\lambda)$ (notation of [17]). Thus, we may assume that $Z(\rho)$ is infinite. Hence, the image of ρ fixes an element x in $\underline{\mathfrak{g}}$, the Lie algebra of $\underline{G}$. Hence, the image of ρ fixes the semi-simple part x_s of x and the nilpotent part x_n of x. If $x_s \neq 0$, then the centralizer of x_s is the Levi subgroup of a proper parabolic and we are done. If $x_n \neq 0$ then the centralizer of x_n is contained in a parabolic subgroup by the following lemma and again we are done. With this the theorem is proved. We owe the next lemma to Robert Steinberg.

Lemma 1.1. Let $\mathfrak{g}$ be a complex semi-simple Lie algebra and $x \in \mathfrak{g}$ a non-zero nilpotent element. The subgroup of G which fixes x in the adjoint action is contained in a proper parabolic subgroup of G.

Proof. By the Jacobson-Morosov Theorem, see Kostant [12], we may choose $h, y \in \mathfrak{g}$ so that $\{x,y,h\}$ is the standard basis for the Lie algebra $s\ell_2(\mathbb{C})$. The element h acts semi-simply on $\mathfrak{g}$ with integer eigenvalues. We may decompose $\mathfrak{g}$ according to $\mathfrak{g} = \oplus\, \mathfrak{g}_i$ where $\mathfrak{g}_i$ is the eigenspace of $\mathfrak{g}$ under h corresponding to the eigenvalue i. We define a parabolic subalgebra $\mathcal{P}$ of $\mathfrak{g}$ by $\mathcal{P} = \oplus_{i \geqq 0} \mathfrak{g}_i$. Let P be the normalizer of $\mathcal{P}$ of $\underline{G}$. By Kostant [12], Theorem 3.6, the element h is unique up to conjugacy by the Lie group G_x corresponding to the subalgebra $\mathfrak{g}_x = \ker \operatorname{ad} x \cap \operatorname{im} \operatorname{ad} x$. Any element of $\ker \operatorname{ad} x$ is a sum of highest weight vectors for h and hence $\mathfrak{g}_x \subset \mathcal{P}$ and $G_x \subset P$. Hence the subgroup P is uniquely determined by x. Hence if g fixes x in the adjoint action then g normalizes P. But a parabolic group is its own normalizer and the lemma is proved.

We now make an assumption that will be satisfied by all pairs $(\Gamma, \underline{G})$ considered in this paper.

Assumption. There exists a stable real representation; that is, there exists a representation $\rho \in R(\Gamma, G)$ such that the image of ρ is not contained in any proper parabolic subgroup of $\underline{G}$.

We now recall that a topological group G is said to act properly on a space X if the map $A: G \times X \to X \times X$ given by $A(g,x) = (gx, x)$ is a proper map.

Proposition 1.1. The actions of $\underline{G}$ on S and G on $S \cap R(\Gamma, G)$ are proper.

The proposition follows from a result in geometric invariant theory. In order to apply this result we have to relate the algebraic geometry definition of properness to the usual one. To this end, we define a morphism of finite type $f: X \to Y$ of affine varieties X and Y to be Zariski proper if it is Zariski universally closed; that is if for every variety Z the morphism $f \times \mathrm{id}: X \times Z \to Y \times Z$ is closed in the Zariski topologies of $X \times Z$ and $Y \times Z$. We now prove a lemma for f as above.

Lemma 1.2. f is proper if and only if f is Zariski proper.

Proof. We leave the implication that proper implies Zariski proper to the reader. Assume f is Zariski proper and let j be an embedding of X in $\mathbb{P}^n$ (the image of X will not be closed in $\mathbb{P}^n$). We have a diagram:

$$\begin{array}{ccc} X & \xrightarrow{j\times f} & \mathbb{P}^n \times Y \\ & {\scriptstyle f}\searrow \quad \swarrow{\scriptstyle p_2} & \\ & Y & \end{array}$$

We claim that f Zariski proper implies $j \times f$ is Zariski proper. Certainly $I \times f : \mathbb{P}^n \times X \to \mathbb{P}^n \times Y$ is Zariski proper. But from the diagram:

$$\begin{array}{ccc} X & \xrightarrow{j\times I} & \mathbb{P}^n \times X \\ & {\scriptstyle j \times f}\searrow & \downarrow{\scriptstyle I \times f} \\ & & \mathbb{P}^n \times Y \end{array}$$

we see that it is enough to prove that $j \times I$ is Zariski proper. But this map is (up to an exchange of factors) the graph map Γ_j of j given by $\Gamma_j(x) = (x, j(x))$. But the graph map of any morphism $h: M \to N$ is proper for it is a closed immersion - the image of Γ_h is the subset of $M \times N$ defined by the equations $h \circ p_1 = p_2$. With this the claim is established. Hence $(j \times f)(X)$ is Zariski closed and hence strongly closed in $\mathbb{P}^n \times Y$. The lemma now follows.

We may now deduce the proposition from the results in Mumford-Fogarty [17], Chapter 2, as follows. By Proposition 2.4, it is sufficient to prove that the action of every one-parameter subgroup $\lambda: \mathbb{C}_m \to \underline{G}$ on S is proper. Suppose this is false. Then there is a sequence $\{\rho_n\}$ contained in a bounded subset of S, a sequence $\{a_n\}$ in $\mathbb{C}^*$ such that $\lim_{n\to\infty} a_n = \infty$ and a one-parameter group μ such that $\lim_{n\to\infty} \mathrm{Ad}\, \mu(a_n) \cdot \rho_n = \rho$ with $\rho \in S$. But the argument in [17], Proposition 2.6, shows that the image of ρ is contained in a proper parabolic subgroup of $\underline{G}$, a contradiction. We note that since G is closed in $\underline{G}$ and $S \cap R(\Gamma, G)$ is closed in S the first statement of the proposition implies the second. With this the proposition is proved.

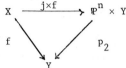

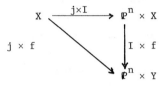

We now prove a technical theorem which will be of use later. See Borel-Wallach [6], page 277, for the definition of a slice to a group action through a point.

Theorem 1.2. The actions of $\underline{G}$ *on* S *and* G *on* $S \cap R(\Gamma,G)$ *admit analytic slices through any* ρ.

Proof. It is sufficient to prove the existence of slices on the stable representations in $R(\Gamma_N,\underline{G})$ and $R(\Gamma_N,G)$ since a slice in $R(\Gamma_N,\underline{G})$ intersects $R(\Gamma,\underline{G})$ in a slice. But $S(\Gamma_N)$ and $S(\Gamma_N) \cap R(\Gamma_N,G)$ are manifolds upon which G acts properly with finite isotropy groups. The theorem now follows from Palais [20].

Corollary. If x_t *is a germ of a curve through* $\pi(\rho)$ *in* $X(\Gamma,\underline{G})$ *or* $X(\Gamma,G)$ *and* ρ *is good then there is a germ of a curve* ρ_t *through* ρ *in* $R(\Gamma,\underline{G})$ *or* $R(\Gamma,G)$ *with image* x_t.

Remark. If ρ is good the quotient map $\underline{\pi}$ induces an analytic equivalence between a neighborhood of ρ in a slice through ρ and a neighborhood of $\underline{\pi}(\rho)$ in $X(\Gamma,\underline{G})$.

We prove another general result concerning $R(\Gamma,\underline{G})$. Let $S^* = \{\rho \in S: Z(\rho) = Z_{\underline{G}}\}$ where $Z_{\underline{G}}$ is the center of $\underline{G}$. S^* is the set of good representations.

Proposition 1.3. S^* *is Zariski open in* $R(\Gamma,\underline{G})$.

Again it is sufficient to prove the proposition for $\Gamma = \Gamma_N$, the free group on N generators. In this case $R(\Gamma,\underline{G})$ is irreducible and consequently S is irreducible.

Lemma 1.3. S^* *is open in the strong topology on* $R(\Gamma,\underline{G})$.

Proof. This is an immediate consequence of the existence of local slices since all identifications on a slice through ρ are made by $Z(\rho)$.

Lemma 1.4. S^* *is constructible.*

Proof. Let Δ be the diagonal in $S \times S$ and $p_2:\underline{G} \times S \to S$ the projection. Then $S - S^* = p_2(A^{-1}(\Delta) - Z_{\underline{G}} \times S)$ is a constructible set.

Since S^* is constructible and strongly open it is Zariski open and the proposition is proved.

Now let Γ be a torsion-free group, $\underline{H}$ a classical simple algebraic group defined over $\mathbb{R}$ with real points H and $\rho_0:\Gamma \to H$ an embedding of Γ into H as a uniform discrete subgroup. If H is not locally isomorphic to $PSL_2(\mathbb{R})$ then Γ is rigid by the Mostow Rigidity Theorem. However, we now suppose that $\underline{H}$ is represented in another algebraic group $\underline{G}$ also defined over $\mathbb{R}$ so that the representation $\underline{H} \to \underline{G}$ is defined over $\mathbb{R}$. We assume the image of $\underline{H}$ is not contained in a proper parabolic subgroup of $\underline{G}$. Then ρ_0 composed with the representation $H \to G$ of real points embeds Γ into G as a discrete subgroup (usually no longer a lattice). We abuse notation and denote the resulting embedding by ρ_0. We now discuss two problems motivated by the classical theory of quasi-Fuchsian groups.

The first problem is to study the deformation space $\mathrm{Hom}(\Gamma,G)/G$. The second is to realize this space as the target of the holonomy mapping of a space of locally homogeneous structures on the original compact locally symmetric space $M = \rho_0(\Gamma)\backslash H/K$ where K is a maximal compact subgroup of H. In the classical case of quasi-Fuchsian groups we have $H = PSL_2(\mathbb{R})$ and $G = PSL_2(\mathbb{C})$. In this case M is a compact surface and the deformation space is the space of holonomy representations of flat conformal structures on M.

Unfortunately, in the general case, the possibilities are severely limited. Recall that a representation $\rho \in \mathrm{Hom}(\Gamma,G)$ is said to be locally rigid if the orbit of ρ in $\mathrm{Hom}(\Gamma,G)$ is open in $\mathrm{Hom}(\Gamma,G)$. We recall that two Lie groups are said to be locally isomorphic if they have isomorphic Lie algebras.

Theorem 1.3. ρ_0 is locally rigid in G unless H is locally isomorphic to $SO(n,1)$ or $SU(n,1)$.

Proof. Assume H is not one of the above groups. Let $\underline{h}$ be the complexification of the Lie algebra of H and $\underline{g}$ the complexification of the Lie algebra of G. By a theorem of Weil, see Raghunathan [21], Theorem 6.7, it is sufficient to prove $H^1(\Gamma,\underline{g}) = 0$. Now, decompose $\underline{g}$ into a sum of irreducible $\underline{H}$ modules according to $\underline{g} = \oplus_{j=1}^m V_j$. Then since $\Gamma \subset H$ we have $H^1(\Gamma,\underline{g}) = \oplus_{j=1}^m H^1(\Gamma,V_j)$. Now $\rho_0(\Gamma)$ is a uniform discrete subgroup of H and V_j is an irreducible representation of H. Hence, by Raghunathan [22], we have $H^1(\Gamma,V_j) = 0$ unless H is as above. With this the theorem is proved.

In fact, Raghunathan's theorem tells us that ρ_0 is locally rigid unless V_j is a symmetric power of the standard representation in case $G = SU(n,1)$ or a space of (Minkowski) spherical harmonics in case $G = SO(n,1)$. We give three examples of representations $H \to G$ for which local rigidity does not follow from Raghunathan's theorem. In each case, let V denote the standard representation of H. We assume $n > 2$ for $SO(n,1)$ and $n > 1$ for $SU(n,1)$ so we have $H^1(\Gamma,\underline{h}) = 0$ by a theorem of Weil, see Raghunathan [21], Chapter VII, section 5. In the orthogonal case, we let S_0^2V denote the (Minkowski) spherical harmonics of degree 2; that is, the "traceless" symmetric 2-tensors. Here the "trace" is the inner product with the $SO(n,1)$ invariant bilinear form $(,)$ using the form on the symmetric 2-tensors induced by $(,)$.

H	G	$\mathfrak{g}$	infinitesimal deformations
$SO(n,1)$	$SO(n+1,1)$	$\mathfrak{h} \oplus V$	$H^1(\Gamma,V)$
$SO(n,1)$	$PGL_{n+1}(\mathbb{R})$	$\mathfrak{h} \oplus V$	$H^1(\Gamma,S_0^2V)$
$SU(n,1)$	$SU(n+1,1)$	$\mathfrak{h} \oplus V$	$H^1(\Gamma,V)$

We will discuss the first and second examples in detail in this paper. We note in the first example $\mathfrak{g}$ may be identified with $\Lambda^2(V \oplus L)$ where $V \oplus L$ is the standard representation of $SO(n+1,1)$ and L is a line invariant under $SO(n,1)$. As representation spaces for $SO(n,1)$ we have:

$$\Lambda^2(V \oplus L) = \Lambda^2V \oplus (L \otimes V) = \Lambda^2V \oplus V.$$

We will use this identification extensively in Sections 6 and 7. In the second case we have $s\ell_{n+1} \approx \Lambda^2V \oplus S_0^2V$ as $SO(n,1)$ modules. In the first case ρ_0 is good if and only if n is even, in the second case ρ_0 is good for all n.

In the third case a more subtle rigidity theorem holds and there are no interesting deformations - see Goldman-Millson [28].

2. Infinitesimal Deformations and Obstructions.

In this section, we review standard material concerning infinitesimal deformations. We begin by recalling the definitions of Eilenberg-MacLane 1-cocycles and coboundaries.

Let V be a vector space and $\rho:\Gamma \to \operatorname{Aut} V$ a representation. Then a 1-cocycle on Γ with coefficients in ρ is a map $c:\Gamma \to V$ such that for $\gamma,\delta \in \Gamma$ we have:

$$c(\gamma\delta) = c(\gamma) + \rho(\gamma)\cdot c(\delta).$$

We let $Z^1(\Gamma,V)$ denote the space of 1-cocycles on Γ with values in V. Elements of $Z^1(\Gamma,V)$ are often called crossed-homomorphisms (with values in V).

A 1-cocycle c is said to be a 1-coboundary if there exists $v \in V$ such that:

$$c(\gamma) = \rho(\gamma)v - v \quad \text{for all} \quad \gamma \in \Gamma$$

We denote the subspace of 1-coboundaries by $B^1(\Gamma,V)$ and define the first cohomology group of Γ with values in V by:

$$H^1(\Gamma,V) = \frac{Z^1(\Gamma,V)}{B^1(\Gamma,V)} .$$

There are similar but more complicated definitions for $Z^p(\Gamma,V)$, $B^p(\Gamma,V)$ and $H^p(\Gamma,V)$ for all $p \geq 1$, see Eilenberg-MacLane [9].

Let X be a real algebraic set in $\mathbb{R}^n$ and $x \in X$. Let $\alpha:(-\varepsilon,\varepsilon) \to X$ be a real analytic curve such that $\alpha(0) = x$. Let $\alpha(t) = \Sigma_{k=0}^{\infty}\alpha_k t^k$ be the Taylor series for α about $t = 0$. We define the <u>leading coefficient</u> of α at $t = 0$ to be α_n if $n > 0$, $\alpha_n \neq 0$ and $\alpha_m = 0$ for $0 < m < n$. We then define the <u>tangent cone</u> TC_x of X at x to be the set of all leading coefficients of curves α as above. If X is smooth at x then TC_x coincides with the tangent space to X at x.

Now let $\rho:\Gamma \to G$ be a representation of Γ into the real points G of an algebraic group $\underline{G}$ defined over $\mathbb{R}$. Let ρ_t be a curve in $\mathrm{Hom}(\Gamma,G)$ with $\rho_0 = \rho$. Let $\dot{\rho}(\gamma) \in T_{\rho(\gamma)}(G)$ be the leading coefficient at $t = 0$ to the curve $\rho_t(\gamma)$ in G. Define a function c from Γ to the Lie algebra $\mathfrak{g}$ of G by:

$$c(\gamma) = \dot{\rho}(\gamma)\rho(\gamma)^{-1}.$$

The following lemma is immediate, observe that Γ acts on $\mathfrak{g}$ by the composition of ρ with the adjoint action of G on $\mathfrak{g}$.

<u>Lemma 2.1</u>. c <u>is a cocycle</u>.

One obtains in this way an embedding of the tangent cone at ρ to $\mathrm{Hom}(\Gamma,G)$ into $Z^1(\Gamma,\mathfrak{g})$. For this reason, the space $Z^1(\Gamma,\mathfrak{g})$ will be called the space of infinitesimal deformations of ρ. We let TC_ρ denote the tangent cone to $\mathrm{Hom}(\Gamma,G)$ at ρ.

Suppose now that ρ_t is a trivial deformation of ρ; that is, suppose there exists a curve g_t in G with $g_0 = 1$, the identity in $\underline{G}$, such that $\rho_t = \mathrm{Ad}\, g_t \cdot \rho$. Let $\dot{g}$ be the tangent vector to g_t at $t = 0$; hence $\dot{g} \in \mathfrak{g}$. Upon differentiating we obtain:

$$c(\gamma) = \dot{g} - \mathrm{Ad}\, \rho(\gamma)\dot{g}$$

and we have proved the following lemma.

Lemma 2.2. *If* c *is tangent to a trivial deformation then* c *is a 1-coboundary. Conversely, every 1-coboundary is a tangent to a trivial deformation.*

Corollary. *If* $c \in TC_\rho$ *then* $c + b \in TC_\rho$ *for all* $b \in B^1(\Gamma,\mathfrak{g})$.

Proof. $c + b$ is tangent to the deformation $\mathrm{Ad}\, g_t \cdot \rho_t$.

Remark. By the previous lemma, the map $d\pi$ annihilates $B^1(\Gamma,\mathfrak{g})$ and induces a map $\overline{d\pi}$ from the image of TC_ρ in $H^1(\Gamma,\mathfrak{g})$ to the tangent cone to $X(\Gamma,G)$ at $\pi(\rho)$. We can obtain more information in case ρ is a good representation. In this case $\pi|S^*(\Gamma_N)$ is a principal bundle so $\overline{d\pi}|H^1(\Gamma,\mathfrak{g})$ is injective. In this case we may identify $d\pi$ with the projection $Z^1(\Gamma,\mathfrak{g})$ to $H^1(\Gamma,\mathfrak{g})$ restricted to TC_ρ.

Lemma 2.3. *If* ρ *is a good representation then* $d\pi$ *maps* TC_ρ *onto the tangent cone of* $X(\Gamma,G)$ *at* $\pi(\rho)$.

Proof. If z is an element in the tangent cone to $X(\Gamma,G)$ at $\pi(\rho)$ then there exists a curve x_t in $X(\Gamma,G)$ with tangent vector z at $t = 0$. But by the corollary to Theorem 1.1 we can lift x_t near $t = 0$ to a curve in $\mathrm{Hom}(\Gamma,G)$. The surjectivity of $d\pi$ follows.

Remark. We call an element of $H^1(\Gamma,\mathfrak{g})$ an infinitesimal deformation of $\pi(\rho)$.

We now derive a necessary condition for an element $c \in Z^1(\Gamma,\mathfrak{g})$ to be the leading coefficient at $t = 0$ to a curve ρ_t in $\mathrm{Hom}(\Gamma,G)$. Recall that the cup-square of $c \in Z^1(\Gamma,\mathfrak{g})$ is the element $[c,c] \in Z^2(\Gamma,\mathfrak{g})$ defined by:

$$[c,c](\gamma,\delta) = [c(\gamma),\mathrm{Ad}\,\rho(\gamma)c(\delta)].$$

Here $[\,,]$ denotes the bracket operation in $\mathfrak{g}$. The following proposition follows from Lemma 2.4 of Goldman-Millson [28].

Proposition 2.1. (i) *If* c *is an element in* $Z^1(\Gamma,\mathfrak{g})$ *which is the leading coefficient at* $t = 0$ *to a curve* ρ_t *in* $R(\Gamma,G)$ *then* $[c,c]$ *is the zero element in* $H^2(\Gamma,\mathfrak{g})$.

(ii) *If* $z \in H^1(\Gamma,\mathfrak{g})$ *is such that* $d\pi(z)$ *is the leading coefficient at* $t = 0$ *to a curve in* $X(\Gamma,G)$ *and* ρ_0 *is good, then* $[z,z] = 0$ *in* $H^2(\Gamma,\mathfrak{g})$.

The second part of the proposition requires some comment. First, it is standard that the cup-product is a well-defined map from $H^1(\Gamma,\mathfrak{g}) \otimes H^1(\Gamma,\mathfrak{g})$ to $H^2(\Gamma,\mathfrak{g})$. Second, by the corollary to Theorem 1.2, a germ in $X(\Gamma,G)$ with leading coefficient z can be lifted to a germ in $\mathrm{Hom}(\Gamma,G)$ with leading coefficient c, where c is a representative cocycle for z. Then $[c,c]$ represents $[z,z]$ and is a coboundary by (i).

Definition. Given an infinitesimal deformation $z \in H^1(\Gamma,\mathfrak{g})$ the class $[z,z] \in H^2(\Gamma,\mathfrak{g})$ is called the first obstruction to the existence of a deformation tangent to z.

Remark. There is an infinite sequence of obstructions to the existence of a deformation tangent to z. Their construction follows the general scheme of Kodaira-Spencer deformation theory. The second obstruction is an analogue of the Massey product and may be interesting for three manifolds.

3. Quasi-Fuchsian Groups in Hyperbolic n-space.

In this section, we will specialize the considerations of Section 1 to the case $\underline{H} = \underline{SO}(n,1)$ and $\underline{G} = \underline{SO}(n+1,1)$. By $\underline{SO}(n,1)$ we will mean the complex points of the algebraic group of orientation preserving isometries of the quadratic form for $\mathbb{C}^{n+1}$ given by $f(x_1,x_2,\ldots,x_{x+1}) = -x_1^2 + x_2^2 +\cdots+ x_{n+1}^2$. The symbol $SO(n,1)$ will denote the real points of $\underline{SO}(n,1)$. We will embed $\underline{SO}(n,1)$ into $\underline{SO}(n+1,1)$ as the isotropy subgroup of the last standard basis vector.

We let Γ be a torsion free group embedded as a uniform discrete subgroup by $\rho_0:\Gamma \to SO(n,1)$. We assume for convenience that $\rho_0(\Gamma) \subset SO_0(n,1)$, the connected component of the identity. We wish to study the space $\mathrm{Hom}(\Gamma,SO(n+1,1))$, its orbit space $\mathrm{Hom}(\Gamma,SO(n+1,1))/SO(n+1,1)$ and its algebraic geometrical quotient $X(\Gamma,SO(n+1,1))$. For many reasons (among them to describe a nice neighborhood of ρ_0 in the quotient) it is useful to impose a technical condition on the representations considered. We observe any representation ρ of Γ in $SO(n+1,1)$ defines an action of Γ on S^n, the boundary of hyperbolic space $\mathbb{H}^{n+1}$.

Definition. A representation ρ in $\mathrm{Hom}(\Gamma,SO(n+1,1))$ is said to be quasi-Fuchsian if the action of Γ via ρ on S^n is quasi-conformally conjugate to the action via ρ_0.

We will call ρ_0 or any representation conjugate by an element of $SO(n+1,1)$ to ρ_0 a Fuchsian representation. Our terminology is classical in the case $n = 2$. We let $R_n(\Gamma)$ denote the space of all quasi-Fuchsian representations of Γ and $T_n(\Gamma)$ the space of conjugacy classes of quasi-Fuchsian representations.

We now prove that the representations in $R_n(\Gamma)$ have several good properties. We note first that for any $\rho \in R_n(\Gamma)$ the subgroup $\rho(\Gamma) \subset SO(n+1,1)$ is discrete since it has a non-empty domain of discontinuity on S^n. Also since the action of any such ρ is topologically conjugate to ρ_0, the group $\rho(\Gamma)$ does not fix any point of S^n and consequently is not contained in any parabolic subgroup of $SO(n+1,1)$. By Morgan [16], Lemma 1.1, we have the following lemma.

Lemma 3.1. *If* $\rho \in R_n(\Gamma)$ *then the* $SO(n+1,1)$ *orbit of* ρ *in* $Hom(\Gamma, SO(n+1,1))$ *is closed*.

Corollary. *If* $\rho \in R_n(\Gamma)$ *then the* $\underline{SO}(n+1,1)$ *orbit of* ρ *in* $Hom(\Gamma, \underline{SO}(n+1,1))$ *is closed*.

Proof. The corollary follows from Birkes [4].

We now show that if $\rho \in R_n(\Gamma)$ and ρ is not Fuchsian then the image of ρ is Zariski dense in $\underline{SO}(n+1,1)$.

By [7], Theorem 4.4.2, we see that it is sufficient to prove that Γ does not leave invariant a totally geodesic subspace of $\mathbb{H}^{n+1}$. But if $\rho(\Gamma)$ leaves invariant a totally geodesic subspace of dimension k with $k < n$, then, since $\rho(\Gamma)$ is discrete, it would operate properly discontinuously on some $\mathbb{H}^k$ and consequently have homological dimension less than or equal to k. But $H_n(\Gamma, R) = \mathbb{R}$. Finally, if $\rho(\Gamma)$ leaves an $\mathbb{H}^n$ invariant then we transform this $\mathbb{H}^n$ to the standard ^{n} by an element of $SO(n+1,1)$. But $M = \rho(\Gamma)\backslash\mathbb{H}^n$ must be compact since $H_n(M,\mathbb{R}) = H_n(\Gamma,\mathbb{R}) = \mathbb{R}$. We can apply Mostow rigidity to conclude ρ is Fuchsian. We obtain:

Lemma 3.2. *A quasi-Fuchsian representation which is not Fuchsian is Zariski dense*.

With these two theorems we have established that $R_n(\Gamma)$ is contained in the subset S of stable representations (Section 1); moreover, if $\rho \in R_n(\Gamma)$ is not Fuchsian then it is good.

The image of S in the variety $X(\Gamma, \underline{SO}(n+1,1))$ is its orbit space - Newstead [19], Proposition 3.8; that is, $\underline{\pi}(\rho_1) = \underline{\pi}(\rho_2)$ if and only if ρ_1 and ρ_2 are conjugate in $\underline{SO}(n+1,1)$.

Lemma 3.3. If $\rho_1, \rho_2 \in Hom(\Gamma, SO(n+1,1))$ and ρ_1 is Zariski dense then $\underline{\pi}(\rho_1) = \underline{\pi}(\rho_2)$ if and only if ρ_1 and ρ_2 are conjugate by an element of SO(n+1,1).

Proof. Since $\underline{\pi}(\rho_1) = \underline{\pi}(\rho_2)$ there exists $g \in \underline{SO}(n+1,1)$ so that $g\rho_1 g^{-1} = \rho_2$. Applying complex conjugation σ we find $\sigma(g)\rho_1\sigma(g)^{-1} = \rho_2$. Hence $\sigma(g)^{-1}g$ centralizes ρ_1. Since ρ_1 is Zariski dense, its centralizer $Z(\rho_1)$ is Z_G, the center of G. Thus either $\sigma(g) = g$ and we are done or $\sigma(g) = -g$. In this second case $g = ih$ with h in $GL_{n+2}(\mathbb{R})$. But we claim that $\underline{SO}(n+1,1)$ contains no pure imaginary matrices (for $n \geq 2$). Indeed h would transform the matrix A of the form f relative the standard basis into its negative. Note ${}^t(ih)A(ih) = A$. But f and $-f$ have different signatures. With this the lemma is proved.

Corollary. $T_n(\Gamma)$ embeds in $X(\Gamma, SO(n+1,1))$.

We are now ready to prove that $R_n(\Gamma)$ is open. First we need a lemma, the main idea of which we owe to John Morgan. We refer the reader to Thurston [25], 8.1 and 8.2, for the definitions and properties of the limit set $\Lambda(\rho(\Gamma))$ (denoted L_Γ in Thurston) and the regular set $\Omega(\rho(\Gamma)) = S^n - \Lambda(\rho(\Gamma))$ for the action of $\rho(\Gamma)$ on S^n.

Let ρ be quasi-Fuchsian. Then we know $\Lambda(\rho(\Gamma))$ is homeomorphic to a sphere and $\Omega(\rho(\Gamma))$ is homeomorphic to two disjoint open hemi-spheres Ω_+ and Ω_-.

Lemma 3.4. $M(\Gamma) = (\mathbb{H}^{n+1} \cup \Omega(\rho(\Gamma)))/\rho(\Gamma)$ is compact.

Proof. $M(\Gamma)$ is a (possibly non-compact) manifold with boundary components the quotients of the two hemi-spheres Ω_+ and Ω_-. Since the action of ρ is topologically conjugate to that of ρ_0, we know $N_+ = \Omega_+/\rho(\Gamma), N_- = \Omega_-/\rho(\Gamma)$ and $\mathbb{H}^n/\rho_0(\Gamma)$ are homeomorphic. Hence N_+ and N_- are compact orientable n-manifolds. Since the universal cover Ω_+ of N_+ embeds into $\mathbb{H}^{n+1} \cup \Omega_+ \cup \Omega_-$ we know $\pi_1(N_+)$ injects into $\pi_1(M(\Gamma))$. But this map is clearly surjective since $\rho_0(\Gamma)$ stabilizes Ω_+. Hence the inclusions $N_+ \subset M(\Gamma)$ and $N_- \subset M(\Gamma)$ are homotopy equivalences. Hence $H_n(M(\Gamma), \mathbb{Z})$ is isomorphic to $\mathbb{Z}$ and the homology classes represented by N_+ and N_- are each generators. Hence N_+ and N_- (with the opposite orientation) are homologous. But then there is a finite chain on $M(\Gamma)$ with boundary $N_+ \cup N_-$ and the lemma is proved.

Corollary. $N(\Gamma) = H(\Lambda(\rho(\Gamma)))/\rho(\Gamma)$ is compact (here $H(X)$ denotes the convex hull of X).

Proof. $N(\Gamma)$ and $M(\Gamma)$ are homotopy equivalent manifolds with boundary, Thurston [25], 8.3.5, via a boundary preserving homotopy equivalence. Hence $N(\Gamma)$ is compact.

We are now ready to prove the main theorem of this section. We will make frequent use of Lok [13], Theorem 2.123 and also use an unpublished argument of Lok.

Theorem 3.1. $R_n(\Gamma)$ is open in $\mathrm{Hom}(\Gamma, SO(n+1,1))$.

Proof. Let ρ be a quasi-Fuchsian representation and α be a small positive number.

Consider the manifold M obtained as the quotient of the open ε-neighborhood of $H(\Lambda(\rho(\Gamma)))$ in $\mathbb{H}^{n+1}$ by $\rho(\Gamma)$. Then ρ is the holonomy of the resulting (incomplete) hyperbolic structure on M. Then by Lok [13], Theorem 2.123, for any $\rho' \in R(\Gamma, SO(n+1,1))$ sufficiently close to ρ there exists an open hyperbolic manifold M' and ψ, a diffeomorphism from M to M' which has the property that ψ maps any geodesic arc in M to an arc in M' of curvature less than α.

We first claim that the developing map $D: \tilde{M}' \to \mathbb{H}^{n+1}$ is injective provided $\alpha < 1$ (here $\tilde{M}'$ denotes the universal cover of M'). Suppose that there exist x', y' in $\tilde{M}'$ so that $D(x') = D(y')$. We can join the preimages of x' and y' under $\tilde{\psi}$ (the lift of ψ) by a geodesic arc since $\tilde{M}$ is convex. Hence x' and y' can be joined by an arc γ of curvature less than 1. But if $D(x') = D(y')$ then $D(\gamma)$ is a closed curve in $\mathbb{H}^{n+1}$ with a single corner and curvature less than 1. No such curve exists in $\mathbb{H}^{n+1}$, see Lok [13] Proposition 2.112.

As a consequence of the result of the previous paragraph we may identify the universal cover of M' with a subset of $\mathbb{H}^{n+1}$ (via the developing map). Of course we may do the same for M. The convex hull of a connected subset X of M' is then defined as the quotient by $\rho'(\Gamma)$ of the convex hull in $\mathbb{H}^{n+1}$ of a connected component of the inverse image of X in M'. Let $\bar{M}' = H^{n+1}/\rho'(\Gamma)$ so $M' \subset \bar{M}'$.

Let $C \subset M$ be the Nielsen convex core of M; that is, $C = H(\Lambda(\rho(\Gamma)))/\rho(\Gamma)$ and let N be the closed ε'-neighborhood of C in M (we assume $\varepsilon' < \varepsilon$). Then N is a strictly convex hyperbolic manifold with C^1 boundary and with holonomy ρ. We claim we can construct a hyperbolic manifold $N' \subset \bar{M}'$ which is diffeomorphic to N, has holonomy ρ' and is strictly convex. We first consider

$\psi(N) \subset M'$. Unfortunately $\psi(N)$ is not necessarily convex but we claim its convex hull is within the $(n+1)\delta(\alpha)$ neighborhood of $\psi(N)$ where $\delta(\alpha) = \cosh^{-1}(1/\sqrt{1-\alpha^2})$. In particular, $\lim_{\alpha\to 0}\delta(\alpha) = 0$. We owe the proof to Larry Lok.

Let p and q be points in $\psi(\tilde{N})$. Then by the argument of the previous paragraph we may join p and q by a curve σ in $\psi(\tilde{N})$ with small curvature α. By Lok [13], Corollary 2.113, the segment σ is homotopic (but not necessarily with endpoints fixed) to a geodesic γ so that σ is within the standard equidistant neighborhood of γ of radius $\delta(\alpha)$. Since this neighborhood is convex, we may find a geodesic γ' joining p and q within this neighborhood. Hence, all geodesic segments in $\mathbb{H}^{n+1}$ joining points of $\psi(\tilde{N})$ lie within the $\delta(\alpha)$ neighborhood of $\psi(\tilde{N})$. We define the k-hull of $\psi(\tilde{N})$ to be the set of convex combinations of k-tuples of points of $\psi(\tilde{N})$. We now show by induction that the k-hull of $\psi(\tilde{N})$ lies within $k\delta(\alpha)$ of $\psi(\tilde{N})$. The previous argument proves the assertion for $k = 1$. Assume that the assertion is proved for the $(k-1)$-hull. We observe that the k-hull is the 1-hull of the $(k-1)$-hull. Let γ be a geodesic segment connecting two points x and y of the $(k-1)$ hull. By the induction hypothesis, there exist points x' and y' in $\psi(\tilde{N})$ so that $d(x',x) < (k-1)\delta(\alpha)$ and $d(y',y) < (k-1)\delta(\alpha)$ - here d denotes the hyperbolic distance. Let γ' be the geodesic segment in $\mathbb{H}^{n+1}$ joining x' and y'. Then, by the case $k = 1$, for any z' on γ' there exists z'' in $\psi(\tilde{N})$ so that $d(z',z'') < \delta(\alpha)$. But the function $d(z,\gamma')$ is a convex function on $\mathbb{H}^{n+1}$ and hence its restriction to γ takes its maximum value at either x or y. Hence, for any z on γ, there exists z' on γ' so that $d(z,z') < (k-1)\delta(\alpha)$. But choosing a z'' as above we find z'' in $\psi(\tilde{N})$ so that $d(z,z'') < k\delta(\alpha)$. We conclude that the k-hull is within $k\delta(\alpha)$ of $\psi(\tilde{N})$. Taking $k = n + 1$ we find that the convex hull of $\psi(\tilde{N})$ is within $(n+1)\delta(\alpha)$ of $\psi(\tilde{N})$.

If we choose α small enough, the boundary of $H(\psi(N))$ will be within a tubular neighborhood of the boundary of $\psi(N)$ and will be transverse to the fibers of that tubular neighborhood. Hence, we may construct a self-diffeomorphism f of $\bar{M}'$ which carries the boundary of $\psi(N)$ to the boundary of $H(\psi(N))$ and consequently carries $\psi(N)$ to $H(\psi(N))$. Now let N' be an ε''-neighborhood of $H(\psi(N))$. Clearly $H(\psi(N))$ and N' are diffeomorphic. Thus, we have found a strictly convex hyperbolic manifold N' with holonomy ρ' and a diffeomorphism

φ from N to N' as required.

We lift φ to a diffeomorphism $\tilde{\varphi}$ from $\tilde{N}$ to $\tilde{N}'$. The sets $\tilde{N}$ and $\tilde{N}'$ are strictly convex submanifolds of $\mathbb{H}^{n+1}$. We may then extend $\tilde{\varphi}$ to $\mathbb{H}^{n+1}$ by mapping normal rays to normal rays as in Thurston [26], 8.3.4, to obtain a quasi-isometry conjugating the action of ρ on $\mathbb{H}^{n+1}$ to the action of ρ' on $\mathbb{H}^{n+1}$. The boundary value of this quasi-isometry gives the required quasi-conformal conjugacy. With this Theorem 3.1 is proved.

Corollary. $T_n(\Gamma)$ *is an open subset of the real quasi-algebraic set determined by the image of* $\mathrm{Hom}(\Gamma, SO(n+1,1))$ *in* $X(\Gamma, SO(n+1,1))$. *We recall that a real quasi-algebraic set is a subset of* $\mathbb{R}^m$ *determined by polynomial equations and inequalities.*

Proof. The image of $\mathrm{Hom}(\Gamma, SO(n+1,1))$ is the image of a real algebraic set by a polynomial mapping and consequently it is quasi-algebraic by the Tarski- Seidenberg Theorem [8].

In order to improve on this result, we must make a more careful study of the real algebraic set $X(\Gamma, SO(n+1,1))$ embedded into $\mathbb{C}^m$ as described in Section 1. We know $T_n(\Gamma)$ is an open subset of the image of $\mathrm{Hom}(\Gamma, SO(n+1,1)))$ under π. We now determine when the image of $\mathrm{Hom}(\Gamma, SO(n+1,1))$ under π is open in $X(\Gamma, SO(n+1,1))$.

Recall that $S^* \subset S$ is the set of representations ρ with the properties:

(i) $\underline{G}\cdot\rho$ is closed in $R(\Gamma, \underline{G})$

(ii) $Z(\rho) = Z_{\underline{G}}$.

We have seen in Section 1 that S^* is Zariski open in $R(\Gamma, \underline{G})$.

We now compute the real points of $S^*/\underline{G}$. Let σ be the conjugation of $\underline{G}$ relative the real form G. We note that the action of σ descends to $X(\Gamma, \underline{SO}(n+1,1))$. A superscript σ on a set will denote the subset of fixed-points for σ. We define an action of τ, the non-trivial element of the Galois group of $\mathbb{C}$ over $\mathbb{R}$, on the regular functions on $R(\Gamma, \underline{SO}(n+1,1))$ by $\tau\cdot f(\rho) = \overline{f(\sigma(\rho))}$ where $\overline{}$ denotes complex conjugation.

Lemma 3.6. $(S/\underline{G})^\sigma$ *is the set of real points* $\underline{\pi}(S)_{\mathbb{R}}$ *of* $\underline{\pi}(S)$.

Proof. By definition $\underline{\pi}(S)_{\mathbb{R}}$ is the set of points in $\underline{\pi}(S)$ where the real invariants take real values. Since the orbits of $\underline{G}$ on S are closed we may identify $S/\underline{G}$ and $\underline{\pi}(S)$. Let f be a real invariant; hence, $\overline{f(\rho)} = f(\sigma(\rho))$. Now $f(\rho)$ is real if and only if $\overline{f(\rho)} = f(\rho)$

or $f(\sigma(\rho)) = f(\rho)$. Since the real invariants separate $\underline{G}$ orbits in S we conclude that ρ and $\sigma(\rho)$ are conjugate under $\underline{G}$ or $\sigma(\underline{\pi}(\rho)) = \underline{\pi}(\rho)$. With this the lemma is proved.

By the previous lemma we know that to compute $\underline{\pi}(S)_{\mathbb{R}}$ we have only to compute the fixed-point set $(S/\underline{G})^{\sigma}$. In fact, it is considerably easier to compute $\underline{\pi}(S^*)_{\mathbb{R}} = (S^*/\underline{G})^{\sigma}$. This we now do.

Let $x \in S^*/\underline{G}$ be a fixed-point of σ. Choose $\rho \in \underline{\pi}^{-1}(x)$. Then $\sigma(\rho) = \mathrm{Ad}\, h\cdot\rho$ for some $h \in \underline{G}$. Applying σ again we find $\sigma(h)h = \pm 1$ since $Z(\rho) = \pm 1$ (in fact if $n+2$ is odd we have $\sigma(h)h = +1$). Hence h is a cocycle, $h \in Z^1(\sigma, \mathrm{Ad}\,\underline{G})$. We recall that h_1 and h_2 in $Z^1(\sigma, \mathrm{Ad}\, G)$ are cohomologous if there exists $g \in \underline{G}$ so that $h_1 = \sigma(g)h_2g^{-1}$. The set of cohomology classes of cocycles is the cohomology set $H^1(\sigma, \mathrm{Ad}\,\underline{G})$. We enumerate this set as $\{h_i\}$ with $i \in I$; in fact, this set is known to be finite. However, this will follow from Lemma 3.7 combined with the fact that a real algebraic set has a finite number of connected components.

For each $h_i \in H^1(\sigma, \mathrm{Ad}\,\underline{G})$ let $(S^*)^{h_i} = \{\rho \in S^* : \sigma(\rho) = h_i\rho h_i^{-1}\}$. Then $(S^*)^{h_i} = (S^*)^{i-h_i}$ and $\underline{\pi}((S^*)^{h_i}) \subset (S^*/\underline{G})^{\sigma}$.

<u>Lemma 3.7.</u> <u>The map</u> $h \to \underline{\pi}((S^*)^h)$ <u>induces a one-to-one correspondence between</u> $H^1(\sigma, \mathrm{Ad}\,\underline{G})$ <u>and the connected components of</u> $(S^*/\underline{G})^{\sigma}$.

<u>Proof</u>. The lemma follows easily since $\mathrm{Ad}\,\underline{G}$ acts freely on S^*.

<u>Corollary</u>. $\pi((S^*)^{\sigma})$ <u>is a connected component of</u> $(S^*/G)^{\sigma}$.

The main result of this section now follows.

<u>Theorem 3.2</u>. (i) $T_n(\Gamma) - \pi(\rho_0)$ is an open subset of the real algebraic set $X(\Gamma, SO(n+1,1))$.

(ii) If n is even, $T_n(\Gamma)$ is an open subset of the real algebraic set $X(\Gamma, SO(n+1,1))$.

<u>Proof</u>. $R_n(\Gamma)$ is open in $\mathrm{Hom}(\Gamma, SO(n+1,1))$ by Theorem 3.1. But by Lemma 3.1 we know $G\cdot\rho_0$ is closed in S^{σ} hence $R_n(\Gamma) - G\rho_0$ is open in S^{σ}. But, by Lemma 3.2, we know $R_n(\Gamma) - G\rho_0 \subset (S^*)^{\sigma}$ and $R_n(\Gamma) - G\rho_0$ is an open G-invariant subset; consequently, its image $T_n(\Gamma) - \pi(\rho_0)$ is open in $\pi((S^*)^{\sigma})$. The statement (i) follows.

In case n is even, an easy calculation shows that for ρ Fuchsian we have $Z(\rho) = Z_{\underline{G}}$. Hence, in this case $R_n(\Gamma) \subset (S^*)^{\sigma}$. With this the theorem is proved.

<u>Remark</u>. In case n is odd, $Z(\rho_0)$ is larger than $Z_{\underline{G}}$ and there is

another component of $(S/\underline{G})^{\sigma}$ containing $\pi(\rho_0)$ namely $\pi(\mathrm{Hom}(\Gamma,SO(n,2))$ where $\underline{SO}(n,2)$ is identified with $\underline{SO}(n+1,1)$ by conjugating by the diagonal matrix with diagonal entries $(1,1,\ldots,1,i)$. Thus, part (ii) of the theorem is false for n odd if $\pi(\mathrm{Hom}(\Gamma,SO(n,2)))$ contains non-trivial curves through ρ_0. The methods of Section 5 show that this is often the case for the standard arithmetic examples.

4. Homology and Cohomology with Local Coefficients and the Crossed Homomorphism Associated to a Hypersurface with Coefficient.

Let X be the underlying space of a simplicial complex and E a flat bundle over X. We wish to define homology and cohomology groups with values in E. We define a p-chain with values in E to be a formal sum $\Sigma_{i=1}^{m} c_i\sigma_i$ where σ_i is an ordered p-simplex and c_i is an element of the fiber of E over the first vertex of σ_i. We denote the group of such chains by $C_p(X,E)$ and define the boundary operator:

$$\partial_p: C_p(X,E) \to C_{p-1}(X,E)$$

by

$$\partial_p(c\sigma) = \langle v_0,v_1\rangle^*(c)\sigma_0 + \sum_{j=1}^{p} (-1)^j c\sigma_j.$$

Here σ_j is the jth face of σ and $\langle v_0,v_1\rangle^*$ is parallel translation from v_0 to v_1 along the edge $\langle v_0,v_1\rangle$. Then $\partial_p^2 = 0$ and we may define homology groups with coefficients in E.

In a similar way cohomology with coefficients in a flat bundle is defined. An E-valued p-cochain on X is a function which assigns to each ordered p-simplex σ an element of the fiber of E over the first vertex of σ. The coboundary $\delta\alpha$ of a p-cochain α is defined on a (p+1)-simplex σ by:

$$\delta\alpha(\sigma) = \langle v_1,v_0\rangle^*\alpha(\sigma_0) + \sum_{j=1}^{p} (-1)^j\alpha(\sigma_j).$$

Here σ_j is the jth face of σ. Then $(\delta)^2 = 0$ and we may define cohomology groups with values in E.

Choose a base-point $x_0 \in X$ and a base-point $\tilde{x}_0$ for $\tilde{X}$, the universal cover of X, such that $\pi(\tilde{x}_0) = x_0$ where $\pi:\tilde{X} \to X$ is the covering. Let E_0 be the fiber of E over x_0, so E_0 is also the fiber of π^*E over $\tilde{x}_0$. Note there is a map $\varepsilon:C^0(\tilde{X},\pi^*E) \to E_0$ using parallel translation of fibers of π^*E to E_0 (this is independent of path on $\tilde{X}$). We use ε to construct a map from $Z^1(X,E)$, the

1-cocycles on X with values in E, to $Z^1(\pi_1(X,x_0),E_0)$, the crossed-homomorphisms on $\pi_1(X,x_0)$ with values in E_0. Let $\alpha \in Z^1(X,E)$ be given. Define $\varphi_\alpha:\pi_1(X,x_0) \to E_0$ on $\gamma \in \pi_1(X,x_0)$ as follows. Let $\tilde{\gamma}$ be a simplicial lift of γ to $\tilde{X}$ starting at x_0. Then:

$$\varphi_\alpha(\gamma) = \varepsilon(\pi^*\alpha(\tilde{\gamma})).$$

It is easily seen that φ_α is a crossed-homomorphism and that the map $\alpha \to \varphi_\alpha$ induces an isomorphism from $H^1(X,E)$ to $H^1(\pi_1(X,x_0),E_0)$.

If α is a p-cochain with coefficients in E and b is a p-chain with coefficients in M and $v:E \otimes M \to N$ is a parallel section of $\mathrm{Hom}(E \otimes M,N)$, then the Kronecker index $\langle\alpha,b\rangle$ is defined by:

$$\langle\alpha,b\rangle = \sum_{\text{simplices } \sigma \text{ in } b} v(\alpha(\sigma) \otimes b_\sigma).$$

Here b_σ is the coefficient of σ in b.

The Kronecker index is well-behaved under ∂ and δ and we get a map:

$$\langle,\rangle:H^p(X,E) \otimes H_p(X,M) \to N.$$

The Kronecker index allows us to identify $H^p(X,E^*)$ and $H_p(X,E)^*$. Indeed, if $\alpha \in H^p(X,E^*)$ and $\langle\alpha,b\rangle = 0$ for all $b \in H_p(X,E)$, then α annihilates the kernel of ∂_p, so α is in the image of δ. This is true because the chain groups are vector spaces so the image of ∂_p is a direct summand.

Now assume X is an oriented n-manifold and E, M, N are flat bundles over X and $v:E \otimes M \to N$ is a parallel section of the flat bundle $\mathrm{Hom}(E \otimes M,N)$. Let $a \in H_{n-p}(X,E)$ and $b \in H_{n-q}(X,M)$ and assume that the simplices of a are in general position with respect to the simplices of b. Then the intersection product $a\cdot b$ of a and b is defined by:

(1) Intersect each simplex σ of a with τ of b to get an $n - (p+q)$ simplex as usual.

(2) Give the resulting simplex the coefficient $v(a_\sigma \otimes b_\tau)$; this has to be given at the initial vertex of the intersection; however, since σ is contractible there is a unique way to move a_σ to any other point of σ and the same for b_τ. In this way we obtain the intersection pairing:

$$H_{n-p}(X,E) \otimes H_{n-q}(X,M) \to H_{n-(p+q)}(X,N)$$

The geometric version of Poincare duality for coefficients

states that the intersection pairing:

$$H_{n-p}(X,E) \otimes H_p(X,E^*) \to \mathbb{R}$$

is a perfect pairing. We then obtain an isomorphism from $H_{n-p}(X,E)$ to $H_p(X,E^*)^*$. Composing this isomorphism with that obtained from the Kronecker index we obtain an isomorphism:

$$PD:H_{n-p}(X,E) \to H^p(X,E).$$

Using v and the usual formula for cup-product of simplicial cochains we obtain a cup-product to be denoted $\cup$:

$$H^p(X,E) \otimes H^q(X,M) \to H^{p+q}(X,N).$$

The cohomological version of Poincare duality tells us that the following pairing is perfect:

$$H^p(X,E) \otimes H^{n-p}(X,E^*) \to \mathbb{R}.$$

Remark. If Z is an $(n-p)$-cycle with coefficients in E, then $PD(Z)$ is characterized by the equation:

$$\langle \eta \cup PD(Z), X\rangle = \langle \eta, Z\rangle$$

for all $\eta \in H^{n-p}(X,E^*)$.

There is a formula relating PD, $\cdot$ and $\cup$ which will be critical to us (in a special case). Let $v:E \otimes M \to N$ be as before. We will not prove the following lemma but we will prove the special case we need, Lemma 4.3.

Lemma 4.1. The following diagram is commutative:

$$\begin{array}{ccc} H^p(X,E) \otimes H^q(X,M) & \overset{\cup}{\to} & H^{p+q}(X,N) \\ PD \otimes PD \uparrow & & PD \uparrow \\ H_{n-p}(X,E) \otimes H_{n-q}(X,M) & \to & H_{n-(p+q)}(X,N) \end{array}$$

There is a particularly simple construction of cycles with coefficients in E. Let Y be a closed, oriented submanifold of X of codimension p and let s be a parallel section of the restriction of E to Y. Let $[Y]$ denote the fundamental cycle of Y so $[Y] = \Sigma_i \sigma_i$, a sum of ordered $n-p$ simplices.

Definition (Notation). $Y \otimes s$ denotes the $(n-p)$ chain with values in E given by $Y \otimes s = \Sigma \sigma_i \otimes s_i$ where s_i is the value of s on the first vertex of σ_i.

Clearly $Y \otimes s$ is a cycle. Suppose now that Y_1 and Y_2 are closed, oriented submanifolds of codimension p and q respectively,

s_1 and s_2 are parallel sections of $E|Y_1$ and $E|Y_2$ respectively, and Y_1 and Y_2 intersect transversely. Then we may simplify the previous general formula defining $(Y_1 \otimes s_1)\cdot(Y_2 \otimes s_2)$ as follows:

(1) Intersect Y_1 and Y_2 in the usual way to obtain a (possibly disconnected) codimension $p+q$ submanifold Z with an intersection multiplicity ± 1 (see Section 7).

(2) Assign to Z the parallel section of $N|Z$ given by $v(s_1, s_2)$.

We wish to relate the intersection product of cycles with coefficients of the previous special type and the cup product of their Poincare duals. We first give a formula for the Poincare dual of a cycle of the type $Y \otimes s$.

Let $U(Y)$ be a tubular neighborhood of Y and φ be a p-cochain with compact support in $U(Y)$ representing the Thom class of $U(Y)$. Let $[U(Y)]$ denote the relative fundamental cycle of $U(Y)$. We extend s to a parallel section of $E|U(Y)$. Then extending $\varphi \otimes s$ by zero we obtain an E-valued p-cocycle on all of Y which we continue to denote $\varphi \otimes s$. We then have the following lemma.

Lemma 4.2. $PD(Y \otimes s) = \varphi \otimes s$.

Proof. Let η be any E^*-valued cocycle. Then, letting $(,)$ denote the pairing between E^* and E we obtain:

$$\begin{aligned}\langle \eta \cup (\varphi \otimes s), X\rangle &= \langle \eta \cup (\varphi \otimes s), [U(Y)]\rangle = \langle (s,\eta) \cup \varphi, [U(Y)]\rangle \\ &= \langle (s,\eta), Y\rangle = \langle \eta, Y \otimes s\rangle .\end{aligned}$$

The next to last inequality follows because (s,η) is a scalar-valued cocycle on $U(Y)$ and φ is the Thom class. The claim now follows from the preceding remark characterizing $PD(Y \otimes s)$.

We can now prove the relation between the intersection product of cycles and the cup product of their duals.

Lemma 4.3. $PD(Z \otimes v(s_1,s_2)) = PD(Y_1 \otimes s_1) \cup PD(Y_2 \otimes s_2)$.

Proof. We choose tubular neighborhoods $U_1(Y_1)$ and $U_2(Y_2)$ and Thom classes φ_1 and φ_2 respectively. But then $\varphi_1 \cup \varphi_2$ represents the Thom class of a suitable tubular neighborhood of $Y_1 \cap Y_2$. We find then

$$\begin{aligned}PD(Y_1 \otimes s_1) \cup PD(Y_2 \otimes s_2) &= (\varphi_1 \otimes s_1) \cup (\varphi_2 \otimes s_2) \\ &= (\varphi_1 \cup \varphi_2) \otimes v(s_1,s_2) \\ &= PD(Z \otimes v(s_1,s_2)).\end{aligned}$$

The last equality follows from Lemma 4.2. The lemma is now proved.

Now suppose Y is an oriented hypersurface in X and $Y \otimes s$ is a cycle with values in E as before. We now give a formula for the crossed homomorphism $\varphi \in H^1(\pi_1(X,x_0),E_0)$ which corresponds to $PD(Y \otimes s)$. Let $\gamma \in \pi_1(X,x_0)$. We suppose γ is transverse to Y. Suppose γ intersects Y at points that are $y_1,y_2,\ldots,y_r$ in γ, that the signs of the intersections at these points $\varepsilon_1,\varepsilon_2,\ldots,\varepsilon_r$ and that $\alpha_i = s(y_i)$ for $i = 1,2,\ldots,r$. Parallel translate α back along γ to x_0 to get α_i'.

<u>Lemma 4.4</u>. $\varphi(\gamma) = \sum_{i=1}^{r} \varepsilon_i \alpha_i'$.

The proof follows immediately from the definition of $\varphi = \varphi_{PD(Y \otimes s)}$ given in the beginning of this section.

There is a decomposition of the fundamental group of X associated to the hypersurface Y. We suppose first that Y separates X into S_1 and S_2. We assume S_2 contains the positive side of Y. Choose a base-point x_1 for X which does not lie on Y and with $x_1 \in S_1$. Let x_2 be a base-point for S_2 and c be a directed arc joining x_1 to x_2 which intersects Y at one point y with multiplicity $+1$. Let a be the segment from x_1 to y and b that from y to x_2, so $c = ab$ in the path groupoid. Let $\{\mu_2,\mu_2,\ldots,\mu_k\}$ be a set of generators for $\pi_1(S_1,x_1)$ and $\{\eta_1,\eta_2,\ldots,\eta_e\}$ be a set of generators for $\pi_1(S_2,x_2)$. Then $\{\mu_1,\ldots,\mu_j,\nu_1,\ldots,\nu_e\}$ is a set of generators for $\pi_1(X,x_1)$ with $\nu_j = c\eta_j c^{-1}$ for $j = 1,2,\ldots,\ell$ by van Kampen's Theorem. Now let $s(y) = \beta$ and $a^*\beta = \alpha$ where a^* denotes parallel translation along a from y to x_0 (a similar notation is used below for other paths). Then by Lemma 4.4 we find for $j = 1,2,\ldots,\ell$:

$$\begin{aligned}\varphi(\nu_j) &= a^*\beta - (c\eta_j b^{-1})^*\beta = \alpha - (c\eta_j b^{-1} a^{-1})^*\alpha \\ &= \alpha - \nu_j^*\alpha = \alpha - \rho(\nu_j)\alpha.\end{aligned}$$

Clearly $\varphi(\mu_j) = 0$ for $j = 1,2,\ldots,k$.

In case $X-Y$ remains connected the $\pi_1(X,x_1)$ has an H·N·N presentation with generators $\{\mu_1,\mu_2,\ldots,\mu_p,\nu\}$ with μ_j in the image of $\pi_1(X - Y,x_1)$ in $\pi_1(X,x_1)$. The extra generator ν meets Y at a single point. An argument similar to the previous one gives:

$$\begin{aligned}\varphi(\mu_j) &= 0 \quad \text{for } j = 1,2,\ldots,p \\ \varphi(\nu) &= \alpha.\end{aligned}$$

We summarize these formulas in a lemma. The notation is as above.

Lemma 4.5. (i) If Y separates then $PD(Y \otimes s)$ corresponds to the unique crossed homomorphism φ given by:

$$\varphi(\mu_j) = 0 \quad \text{for} \quad j = 1,2,\ldots,k$$
$$\varphi(\nu_j) = \alpha - \rho(\nu_j)\alpha.$$

(ii) If Y does not separate then $PD(Y \otimes s)$ corresponds to the unique crossed homomorphism φ given by:

$$\varphi(\mu_j) = 0 \quad \text{for} \quad j = 1,2,\ldots,p$$
$$\varphi(\nu) = \alpha.$$

We will often identify a parallel section s along a submanifold Y containing the base-point x_0 for X with its value α at x_0. Then α is an invariant for $\pi_1(Y,x_0)$. Given such an invariant α, we will often denote the corresponding cycle by $Y \otimes \alpha$.

5. Algebraic Bending and a Lower Bound on the Dimensions of the Deformation Spaces.

In this section, we construct deformations of Γ in the conformal and projective groups corresponding to disjoint, non-singular, two-sided, totally geodesic hypersurfaces. By a two-sided hypersurface we mean one with a trivial normal bundle - it does not necessarily separate M. We begin with the case of a single hypersurface.

Suppose M is a compact manifold and M_1 is an embedded two-sided connected hypersurface in M. We suppose moreover, that we are given a representation $\rho:\pi_1(M) \to G$ where G is a Lie group with Lie algebra $\mathfrak{g}$. We abbreviate $\pi_1(M)$ to Γ and $\pi_1(M_1)$ to A.

Lemma 5.1. Suppose $\rho(\pi_1(M_1))$ has an invariant x_1 in $\mathfrak{g}$ such that x_1 is not invariant under $\rho(\Gamma)$. Then $R(\Gamma,G)$ contains a non-constant curve through ρ.

Proof. We first assume M_1 separates M into 2 parts S_1 and S_2. Then Γ is an amalgam $\pi_1(S_1)*_A\pi_1(S_2)$. The vector x_1 cannot be invariant under both $\rho(\pi_1(S_1))$ and $\rho(\pi_1(S_2))$ since they generate $\rho(\Gamma)$. We assume x_1 is not invariant under $\rho(\pi_1(S_2))$. Then we define a curve ρ_t in $R(\Gamma,G)$ by:

$$\rho_t|\pi_1(S_1) = \rho$$
$$\rho_t|\pi_1(S_2) = \mathrm{Ad}\, R(t)\cdot\rho \quad (\text{where } R(t) = \operatorname{expt} x_1).$$

The representations $\rho_t|\pi_1(S_1)$ and $\rho_t|\pi_1(S_2)$ agree on A; hence, by the universal property of amalgams we obtain a representation of Γ.

We next consider the case in which $S = M - M_1$ remains connected. Then Γ is an H·N·N group, $\Gamma = \pi_1(S)_{*A}$. Hence we have a generator ν of Γ such that the only relations involving ν are of the form $\nu^{-1}j_1(a)\nu = j_2(a)$ where j_1 and j_2 are the inclusions of the fundamental groups of the two sides of M_1 into $\pi_1(S)$. We let $R(t) = \exp tx_1$ where x_1 is invariant under $\rho(j_1(A))$ and we define:

$$\rho_t | \pi_1(S) = \rho$$
$$\rho_t(\nu) = R(t)\rho(\nu).$$

Note $\rho_t(\nu)^{-1}\rho(j_1(a))\rho_t(\nu)$ is constant in t and so we obtain representation $\rho_t : \Gamma \to G$ for all t.

<u>Definition</u>. A trivial deformation of ρ parametrized by a set T is one obtained by conjugating ρ by a family of elements of G parametrized by T.

We may prove the non-triviality of the above deformation by computing the class of the cocycle $c \in Z^1(\Gamma, \mathfrak{g})$ tangent to ρ_t. Since c is a crossed homomorphism, it is determined by its values on a set of generators. We choose generators for $\pi_1(M)$ as described at the end of Section 4.

A straightforward calculation then yields the following lemma.

<u>Lemma 5.2</u>. <u>In case</u> M_1 <u>separates we have</u>:

$$c(\mu) = 0 \qquad \text{for } \mu \in \pi_1(S_1)$$
$$c(\nu) = x - \rho(\nu)x\rho(\nu)^{-1} \qquad \text{for } \nu \in \pi_1(S_2)$$

<u>In case</u> M_1 <u>does not separate we have</u>:

$$c(\mu) = 0 \quad \text{for } \mu \in \pi_1(S)$$
$$c(\nu) = x.$$

Lemma 5.2 together with Lemma 4.5 gives the following theorem of fundamental importance in what follows.

<u>Theorem 5.1</u>. <u>The derivative of the bending deformation</u> ρ_t <u>and the Poincare dual of the cycle with coefficients</u> $M_1 \otimes x$ <u>coincide as elements in</u> $H^1(\Gamma, \mathfrak{g})$.

To show that the deformation ρ_t is non-trivial it is sufficient to prove that c is a non-zero element of $H^1(\Gamma, \mathfrak{g})$. To do this and to treat the general case of r embedded hypersurfaces we introduce the graph associated to a collection of two-sided hypersurfaces of a manifold M.

Let M be a connected n-dimensional manifold and $\tilde{M}$ the universal cover of M. In the next lemma Γ denotes the group of covering transformations of $\pi:\tilde{M} \to M$.

Lemma 5.3. <u>Suppose</u> $M_1, M_2, \ldots, M_r$ <u>are disjoint non-singular two-sided connected hypersurfaces in</u> M. <u>Then there exists an oriented tree</u> X <u>such that</u> Γ <u>acts on</u> X <u>without inversions and such that</u> $Y = X/\Gamma$ <u>has</u> r <u>edges</u>.

Proof. We define the oriented graph Z associated to any collection of disjoint two-sided connected hypersurfaces $\{M_i\}_{i\in I}$ in a manifold M. The vertices of Z are the components $\{S_j\}_{j\in J}$ of $M - \cup_{i\in I} M_i$ and the edges are the M_i's. We choose a tubular neighborhood around M_i with boundary $M_i^+ \amalg M_i^-$. Then, the origin $o(M_i)$ of the edge M_i is the component of $M - \cup_{i\in I} M_i$ containing M_i^- and the terminus $t(M_i)$ of M_i is the component of $M - \cup_{i\in I} M$ containing M_i^+. We observe that if M is the disjoint union of two submanifolds M' and M'' then the graph Z is the disjoint union of two subgraphs (possibly empty) Z' and Z''.

Now, given M as in the statement of the lemma, we have the graph Z attached to the collection $\{M_i\}_{i=1}^r$. We also have the collection of hypersurfaces in $\tilde{M}$ formed from the connected components M'_α of inverse images of the M_i under the covering $\pi:\tilde{M} \to M$. The set $\{M'_\alpha\}$ separates $\tilde{M}$ into regions S'_β. We let $\tilde{Z}$ denote the corresponding graph.

We claim $\tilde{Z}$ is a tree. $\tilde{Z}$ is connected for if S'_β and S'_γ are components we may connect them by a path in $\tilde{M}$ crossing a finite number of hypersurfaces. This gives an edge path between the vertices S'_β and S'_γ in $\tilde{Z}$. To show $\tilde{Z}$ is a tree it is now sufficient to show that the removal of any non-extreme edge from $\tilde{Z}$ disconnects $\tilde{Z}$ (an extreme edge is an edge containing an extreme vertex). But the graph obtained upon removing a corresponding M'_α is the graph associated to the collection of hypersurfaces $\{M'_\beta : \beta \neq \alpha\}$ in the manifold $\tilde{M} - M'_\alpha$. But $\tilde{M} - M'_\alpha$ has two components (since any closed hypersurface in $\tilde{M}$ must separate $\tilde{M}$), each of which contains hypersurfaces from the collection $\{M'_\beta : \beta \neq \alpha\}$. Hence the new graph is the union of two disjoint proper subgraphs by the observation from the first paragraph. With this the claim is established.

Since the collection $\{M'_\alpha\}$ is Γ-invariant, we see that Γ acts on the tree $\tilde{Z}$. Since the M_i's are two-sided, Γ maps the positive and negative sides of M' to the corresponding sides of $\gamma M'_\alpha$. Hence Γ acts on $\tilde{Z}$ without inversions. We take $X = \tilde{Z}$.

We claim that Z is the quotient of $\tilde{Z}$ under the action of Γ. There is a map $p:\tilde{Z} \to Z$ given by sending the vertex on $\tilde{Z}$ corresponding to S'_β to the vertex of Z corresponding to $\pi(S'_\beta)$ and the edge in $\tilde{Z}$ corresponding to M'_α to the edge of Z corresponding to $\pi(M'_\alpha)$. Clearly p is incidence preserving, bijective on Γ-orbits and factors through $\tilde{Z}/\Gamma$. If we take $Y = Z$, the lemma is proved.

Corollary. Choose a maximal tree T in Y. Then Γ is isomorphic to the fundamental group $\pi_1(\Gamma,Y,T)$ (here the notation is as in Serre [23], page 42).

Proof. The corollary is the Bass-Serre Theorem, Serre [23], Theorem 13.

We can now give a sufficient condition for the deformations of Lemma 5.1 to be non-trivial. We first treat the case in which M_1 separates M into two parts S_1 and S_2 with S_2 the positive side. ρ_t will denote the deformation of Lemma 5.1. We abbreviate $\pi_1(M_1)$ to A and $\pi_1(S_1)$ and $\pi_1(S_2)$ to B_1 and B_2 respectively.

Lemma 5.4. Suppose neither $\rho(B_1)$ nor $\rho(B_2)$ has a non-zero invariant in g. Then ρ_t is a non-trivial deformation.

Proof. Since Γ acts on the tree X we have a cohomology exact sequence relating the cohomology of Γ with coefficients in g to that of the stabilizers of the vertices and edges of X (Serre [23], Proposition 13). In particular we have:

$$H^0(B_1,g) \oplus H^0(B_2,g) \to H^0(A,g) \to H^1(\Gamma,g)$$

By hypothesis $H^0(B\ ,g)$ and $H^0(B_2,g)$ are both zero; hence $\delta_*:H^0(A\ \ g) \to H^1(\Gamma,g)$ is injective. But $\delta_* x_1 = -c$ (lift x_1 back to $(0,x_1)$), and $\delta(0,x_1)(\mu) = 0$ and $\delta(0,x_1)(\nu) = (0,x_1 - \rho(\nu)x_1\rho(\nu)^{-1})$. With this the lemma is proved.

We next treat the case in which $M - M_1$ remains connected. We abbreviate $\pi_1(S)$ to B. Again ρ_t denotes the deformation of Lemma 5.1.

Lemma 5.5. Suppose $\rho(B)$ has no non-zero invariant in g. Then ρ_t is a non-trivial deformation.

Proof. The cohomology exact sequence associated to the action of Γ on X now becomes in part:

$$H^0(B,\mathfrak{g}) \to H^0(A,\mathfrak{g}) \to H^1(\Gamma,\mathfrak{g}).$$

Since $H^0(B,\mathfrak{g})$ is zero by hypothesis, we find again that $\delta_*:H^0(A,\mathfrak{g}) \to H^1(\Gamma,\mathfrak{g})$ is injective.

We claim that again we have $\delta_* x_1 = -c$. To prove this claim we consider the diagram of short exact sequences of inhomogeneous cochains: (Serre [23], page 126):

$$\begin{array}{ccccccccc}
0 \to & C^1(\Gamma,\mathfrak{g}) & \to & C^1(\Gamma,\mathrm{ind}_B^\Gamma 1 \otimes \mathfrak{g}) & \to & C^1(\Gamma,\mathrm{ind}_A^\Gamma I \otimes \mathfrak{g}) & \to 0 \\
 & & & \uparrow \delta_1 & & \uparrow & \\
0 \longrightarrow & \mathfrak{g} & \longrightarrow & \mathrm{ind}_B^\Gamma 1 \otimes \mathfrak{g} & \xrightarrow{\delta_2} & \mathrm{ind}_A^\Gamma 1 \otimes \mathfrak{g} & \longrightarrow 0
\end{array}$$

Here $\mathrm{ind}_B^\Gamma 1 \otimes \mathfrak{g}$ means the representation with underlying vector space $f:\Gamma \to \mathfrak{g}:f(\gamma\mu) = f(\gamma),\mu \in B\}$ and Γ action given by:

$$\gamma\cdot f(\eta) = \rho(\gamma)f(\gamma^{-1}\eta)\rho(\gamma)^{-1}.$$

The previous diagram arises from the double complex of Eilenberg-MacLane cochains with values in the $\mathfrak{g}$-valued cellular cochains of X together with Γ-equivariant identifications of $\mathrm{ind}_B^\Gamma 1 \otimes \mathfrak{g}$ with $C^0(X) \otimes \mathfrak{g}$ and $\mathrm{ind}_A^\Gamma 1 \otimes \mathfrak{g}$ with $C^1(X) \otimes \mathfrak{g}$. Under these identifications the cellular coboundary $d:C^0(X) \otimes \mathfrak{g} \to C^1(X) \otimes \mathfrak{g}$ is identified with $\delta_2:\mathrm{ind}_B^\Gamma 1 \otimes \mathfrak{g} \to \mathrm{ind}_A^\Gamma 1 \otimes \mathfrak{g}$ given by:

$$\delta_2 f(\gamma) = f(\gamma\nu) - f(\gamma).$$

We now compute $\delta_* x_1$. Our strategy is to identify $x_1 \in \mathfrak{g}^A$ with an element h in $\mathrm{ind}_A^\Gamma 1 \otimes \mathfrak{g}$ (Frobenius reciprocity), compute a suitable preimage f of h under δ_2, apply δ_1 to f and find a preimage b to $\delta_1 f$ in $C^1(\Gamma,\mathfrak{g})$. The class of b will then by definition be $\delta_* x_1$.

Clearly h is given by the formula $h(\gamma) = \rho(\gamma)x_1\rho(\gamma)^{-1}$. Let $f \in \mathrm{ind}_B^\Gamma 1 \otimes \mathfrak{g}$ satisfy:

(i) $f(1) = 0$

(ii) $\delta_2 f(\gamma) = f(\gamma\nu) - f(\gamma) = h(\gamma)$.

We note that f exists because X is a tree and $\nu^{-1}A\nu \subset B$. Now $\delta_1 f \in C^1(\Gamma,\mathrm{ind}_B^\Gamma 1 \otimes \mathfrak{g})$. We may identify the space of such cochains F with functions $\hat{F}$ from $\Gamma \times \Gamma$ into $\mathfrak{g}$ satisfying:

$$\hat{F}(\gamma_1,\gamma_2\mu) = \hat{F}(\gamma_1,\gamma_2) \quad \text{for } \mu \in B.$$

Then we have $\delta_1^{\wedge} f(\gamma_1,\gamma_2)$ is independent of γ_2 and the cochain $b \in C^1(\Gamma,\mathfrak{g})$ - see above, is defined by $b(\gamma) = \delta_1^{\wedge} f(\gamma,1)$. Now according to the formula for the Eilenberg-MacLane coboundary we have:

$$\delta_1^{\hat{}} f(\gamma_1,\gamma_2) = \rho(\gamma_1)f(\gamma_1^{-1}\gamma_2)\rho(\gamma_1)^{-1} - f(\gamma_2)$$

and

$$b(\gamma) = \rho(\gamma)f(\gamma^{-1})\rho(\gamma)^{-1} - f(1) = \rho(\gamma)f(\gamma^{-1})\rho(\gamma)^{-1}.$$

We claim $b = -c$. We have only to check that they coincide on B and on ν since they are both crossed homomorphisms.

If $\mu \in B$ then:

$$b(\mu) = \rho(\mu)f(\mu^{-1})\rho(\mu)^{-1} = \rho(\mu)f(1)\rho(\mu)^{-1} = 0 = -c(\mu).$$

Before evaluating b on ν we note that as a consequence of (ii) with $\gamma = \nu^{-1}$ we have:

$$f(1) - f(\nu^{-1}) = \rho(\nu^{-1})x_1\rho(\nu) \quad \text{or} \quad f(\nu^{-1}) = -\rho(\nu^{-1})x_1\rho(\nu)$$

and

$$\rho(\nu)f(\nu^{-1})\rho(\nu)^{-1} = -x_1.$$

Finally then we obtain:

$$b(\nu) = \rho(\nu)f(\nu^{-1})\rho(\nu)^{-1} = -x_1 = -c(\nu).$$

With this the lemma is proved.

We now consider the general case in which M is a compact manifold containing r disjoint two-sided embedded connected hypersurfaces $M_1,M_2,\ldots,M_r$. We suppose we have a representation $\rho:\Gamma \to G$ where G is the set of real points of a reductive algebraic group $\underline{G}$ defined over $\mathbb{R}$. In what follows we will consider a graph Y with r edges containing a maximal tree T with b edges. Our strategy is to construct a deformation of the group associated to T.

Lemma 5.6. *Suppose ρ is a representation into G of the fundamental group of a tree T of groups such that every edge group has a non-zero invariant in $\mathfrak{g}$. Suppose T has b edges and P is a vertex of T. Then there exists a b-parameter family of deformations of ρ which is constant on Γ_P and is trivial when restricted to any vertex group.*

Proof. For each integer n we let T_n be the set of vertices at distance n from P. If $Q \in T_n$, with $n \geq 1$, there is a single vertex Q' at distance strictly less than n from Q to which Q is adjacent. The correspondence $Q \to Q'$ defines a map of T_n into T_{n-1}. If $Q \in T_n$ then we call the vertices in $\bigcup_{m>n} f_m^{-1}(Q)$ the predecssors of Q and the vertices of $f_{n+1}^{-1}(Q)$ the immediate predecessors of Q. The vertex Q together with its predecessors form the vertex set of a subtree of T.

Let Q_1 be an immediate predecessor of P and e_1 be the edge joining P and Q_1. Let A_1 be the edge group associated to A_1 and x_1 an invariant of $\rho(A_1)$ in g. Let $R(t) = \exp tx_1$. Then we define a 1-parameter family ρ_t of representations of Γ by defining:

$$\rho_t|\Gamma_Q = \mathrm{Ad}\, R(t)\rho|\Gamma_Q \quad \text{if } Q = Q_1 \text{ or } Q \text{ is a predecessor of } Q_1$$
$$\rho_t|\Gamma_Q = \rho|\Gamma_Q \quad \text{otherwise.}$$

As Q_1 varies through T_1, we obtain a b_1-parameter family ρ_{t_1}, for $t_1 \in \mathbb{R}^{b_1}$, of deformations of ρ. Here b_1 is the number of immediate predecessors of P. Clearly the family ρ_{t_1} satisfies all the hypotheses of the lemma.

Now we choose a vertex $Q_1 \in T_1$ and let Q_2 be an immediate predecessor of Q_1. Let e_2 be the edge joining Q_1 and Q_2. Let A_2 be the edge group associated to e_2. Choose $t_1 \in \mathbb{R}^{b_1}$. Then the representation $\rho_{t_1}|A_2$ has a non-zero invariant $x_2 = x_2(t_1)$ in g because it is conjugate to $\rho|A_2$. Moreover we may choose x_2 to be an analytic function of t_1. Let $R(t_1,t) = \exp tx_2(t_1)$. We define a (b_1+1)-parameter family $\rho_{t_1,t}$ of deformations of ρ by defining:

$$\rho_{t_1,t}|\Gamma_Q = \mathrm{Ad}\, R(t_1,t)\cdot\rho_{t_1}|\Gamma_Q \quad \text{if } Q = Q_2 \text{ or } Q \text{ is a predecessor of } Q_2$$
$$\rho_{t_1,t}|\Gamma_Q = \rho_{t_1}|\Gamma_Q \quad \text{otherwise.}$$

Continuing in this way we obtain the lemma.

Lemma 5.7. Suppose ρ is a representation of the fundamental group of a graph of groups $\pi_1(\Gamma,Y,T)$ into G so that every edge group has a non-zero invariant in g. Suppose that Y has r edges and P is a base vertex. Then there exists an r-parameter family of deformations of ρ which is constant on Γ_P and is a trivial deformation of ρ restricted to any vertex group.

Proof. Let Ω be the subgroup of Γ generated by the vertex groups. Then Ω is the fundamental group of the subgraph of groups corresponding to T and we may apply Lemma 5.6 to obtain a b-parameter family ρ_t of deformations of $\rho|\Omega$ satisfying the hypotheses of Lemma 5.6.

We claim we may extend ρ_t to a b-parameter family of representations $\tilde{\rho}_t$ of Γ. It remains to extend ρ_t to the generators v corresponding to the edges e of Y which are not in T. There are two cases. We assume first that e is a loop and let Q be the origin (and terminus) of e. Then the relations involving ν in $\pi_1(\Gamma,Y,T)$ are of the form:

$$\nu j_2(a)\nu^{-1} = j_1(a) \quad \text{for} \quad a \in A$$

where A is the edge group associated to e and j_1 and j_2 are two embeddings of A into Γ_Q. Since $\rho_t|\Gamma_Q$ is trivial there is a b-parameter family R_t of elements of G so that $\rho_t|\Gamma_Q = \mathrm{Ad}\, R_t \cdot \rho|\Gamma_Q$. We define $\tilde{\rho}_t(\nu) = \mathrm{Ad}\, R_t \cdot \rho(\nu)$. Then the relations involving ν are satisfied.

We now suppose that e is not a loop. Let P and Q be the origin and terminus of e. Then the relations involving ν are of the form $\nu j_2(a)\nu^{-1} = j_1(a)$ where j_1 is the inclusion of A into Γ_P and j_2 is the inclusion of A into Γ_Q. Now there exist b-parameter families R_t' and R_t of elements of G so that $\rho_t|\Gamma_{P_1} = \mathrm{Ad}\, R_t' \cdot \rho|\Gamma_P$ and $\rho_t|\Gamma_Q = \mathrm{Ad}\, R_t \cdot \rho|\Gamma_Q$. Define $\tilde{\rho}_t(\nu) = R_t' \rho(\nu) R_t^{-1}$. Then the relations involving ν are satisfied and we have proved the claim.

We now extend $\tilde{\rho}_t$ to an r-parameter family $\rho_{t,u}$ of deformations of ρ for $u = (u_1,u_2,\ldots,u_\ell)$ with $\ell = r - b$.

We define $\rho_{t,u}$ on the generators of $\pi_1(\Gamma,Y,T)$. If $\gamma \in \Omega$, then $\rho_{t,u}(\gamma) = \tilde{\rho}_t(\gamma)$. Consider the generator ν_j associated to an edge e_j of $Y - T$. Either e_j is a loop with vertex Q or it is an edge with origin Q and terminus Q'. If A_j is the edge group associated to e_j then we have the embedding $j_1 : A_j \to \Gamma_Q$. Now $\tilde{\rho}_t|j_1(A_j)$ is trivial; hence $\tilde{\rho}_t|j_1(A_j)$ admits a non-zero invariant x_j' in $\mathfrak{g}$. Put $R(u_j) = \exp u_j x_j'$ and define $\rho_{t,u}(\nu_j) = R(u_j)\tilde{\rho}_j(\nu_j)$. With this the lemma is proved.

We now compute the derivative of $\rho_{t,u}$. We use the exact cohomology sequence obtained from the action of Γ on X, Serre [23], Proposition 13. We require some more notation. For $j = 1,2,\ldots,r$, let A_j be the edge group corresponding to the edge e_j of Y. We assume the edges are ordered so that $e_1,e_2,\ldots,e_b$ are in T. We enumerate the vertices of Y as $P = P_1,P_2,\ldots,P_m$ so that the vertices of T_n (see Lemma 5.6) come before those of Y_m for $m > n$. We orient the edges of Y so that $o(e)$ comes before $t(e)$ in the enumeration. Let B_k for $k = 1,2,\ldots,m$ be the corresponding vertex groups. Then we have the sequence (exact at the middle):

$$\bigoplus_{k=1}^{m} H^0(B_k,\mathfrak{g}) \to \bigoplus_{j=1}^{r} H^0(A_j,\mathfrak{g}) \to H^1(\Gamma,\mathfrak{g}) \qquad (*)$$

In Lemma 5.7 we constructed an analytic map $\Phi(t,u):\mathbb{R}^r \to R(\Gamma,G) \subset G^N$ given by $\Phi(t,y) = \rho_{t,u}$.

Lemma 5.8. Assume that no vertex group has a non-zero invariant in g (under ρ). Then the differential of Φ has rank r at the origin of $\mathbb{R}^r$.

Proof. We remind the reader of the enumeration of the vertices of Y in the previous paragraph. Let $\{\gamma_{ki}:i \in I_k\}$ be a set of generators for Γ_{P_k}. We let $\nu_1,\nu_2,\ldots,\nu_\ell$ be the generators for $\pi_1(\Gamma,Y,T)$ corresponding to the positively oriented edges of Y that are not in T. We have assumed that there are N generators for Γ in all. This choice of generators gives an embedding of $R(\Gamma,G)$ into G^N by:

$$\rho \to (\rho(\gamma_1),\ldots,\rho(\gamma_m),\rho(\nu_1),\ldots,\rho(\nu_\ell)).$$

Here we have abbreviated the coordinates corresponding to the generators $\{\gamma_{ki}:i \in I_K\}$ by a single symbol $\rho(\gamma_k)$.

We may then consider Φ as a map from $\mathbb{R}^r$ into G^N. It is convenient to define $\Phi_1:\mathbb{R}^r \to G^m$ and $\Phi_2:\mathbb{R}^r \to G$ by:

$$\Phi_1(t,u) = (\rho_{t,u}(\gamma_1),\ldots,\rho_{t,u}(\gamma_m))$$
$$\Phi_2(t,u) = (\rho_{t,u}(\nu_1),\ldots,\rho_{t,u}(\nu_\ell))$$

Then $\Phi(t,u) = (\Phi_1(t,u),\Phi_2(t,u))$ and $\Phi_1(t,u)$ does not depend on u. Thus to prove the lemma it is sufficient to prove that $D_t\Phi_1(0,0)$ has rank b and $D_u\Phi_2(0,0)$ has rank ℓ. Now $D_t\Phi_1(0,0)$ and $D_u\Phi_2(0,0)$ take values in $Z^1(\Gamma,g)$. We denote their compositions with the projection to $H^1(\Gamma,g)$ by $\overline{D_t\Phi_1}$ and $\overline{D_u\Phi_2}$. It is sufficient to prove these latter two maps have ranks b and ℓ.

We first compute $\overline{D_u\Phi_2}$. From Lemma 5.7 we have:

$$\Phi_2(0,u) = (R(u_1)\rho(s_1),\ldots,R(u_\ell)\rho(s_\ell)).$$

But the calculation of $\partial\Phi_2/\partial u_i(0,0)$ is identical to that of Lemma 5.5. We find a commutative diagram:

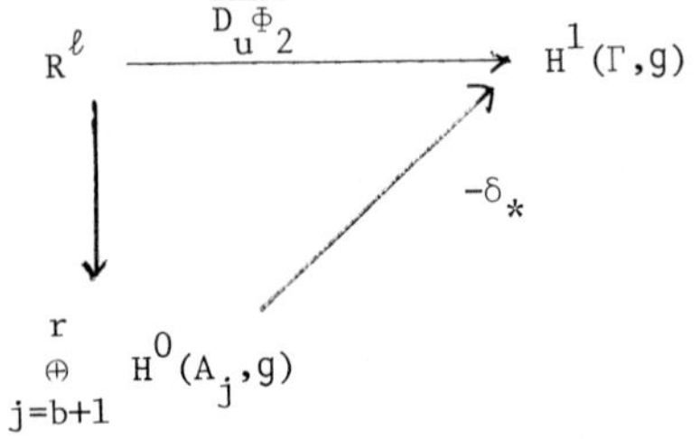

Here the vertical arrow is an isomorphism mapping $(u_1, u_2, \ldots, u_\ell)$ to $(u_1 x_1, \ldots, u_\ell x_\ell)$. Since $H^0(B_j, g)$ is zero for all j it follows from the exactness of (*) that δ_* is injective and consequently $\overline{D_u \Phi_2}$ has rank ℓ.

We now compute $\overline{D_t \Phi_1}$. We claim that again we have a commutative diagram:

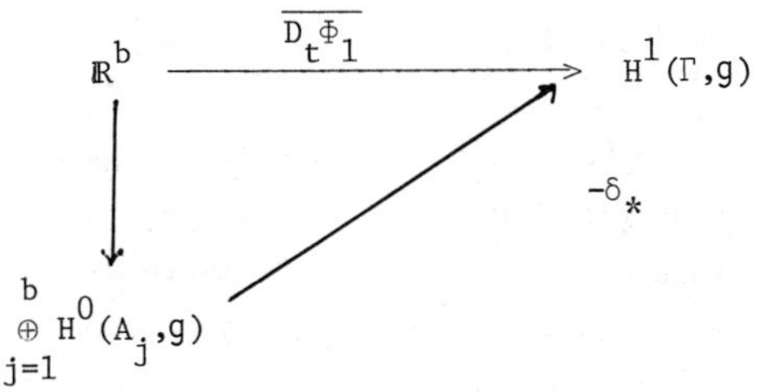

Here the vertical map sends $(t_1, t_2, \ldots, t_b)$ to $(t_1 x_1, \ldots, t_b x_b)$. The claim is equivalent to the formula:

$$\frac{\partial \Phi}{\partial t_j}(0,0) = -\delta_*(0, \ldots, x_j, 0, \ldots, 0) \text{ where } x_j \text{ is in the jth component.}$$

By the construction of Lemma 5.6 there exists an edge e with origin Q and terminus Q' and edge group A so that x_j is invariant under $\rho(A)$ and Φ satisfies:

$$\Phi((0, \ldots, t_j, \ldots, 0), (0, \ldots, 0)) = (\rho_{t_j}(\gamma_1), \ldots, \rho_{t_j}(\gamma_m))$$

where:

$$\rho_{t_j}(\gamma_k) = R(t_j)\rho(\gamma_k)R(t_j)^{-1} \quad \text{if } P_k \text{ is } Q' \text{ or a predecessor of } Q'$$
$$\rho_{t_j}(\gamma_k) = \rho(\gamma_k) \quad \text{otherwise.}$$

Hence $\partial\Phi/\partial t_j(0,0)$ is identified with the cocycle c_j given by:

$$c_j(\gamma_k) = x_j - \rho(\gamma_k) x_j \rho(\gamma_k)^{-1} \quad \text{if } P_k \text{ is } Q' \text{ or a predecessor of } Q'$$
$$c_j(\gamma_k) = 0 \quad \text{otherwise.}$$

But to compute $\delta_*(0, \ldots, x_j, \ldots, 0)$ we observe that an inverse image of $(0, \ldots, x_j, \ldots, 0)$ in $\oplus_{k=1}^{m} H^0(B_k, g)$ is given by $\alpha = (\alpha_k)$, where α is given by:

$$\alpha_k = x_j \quad \text{if } P_k \text{ is } Q' \text{ or a predecessor of } Q'$$
$$\alpha_k = 0 \quad \text{otherwise.}$$

Clearly $\delta\alpha = -c_j$ and the lemma is proved (here δ is the Eilenberg-MacLane coboundary).

Remark. In the course of the proof we have proved the following diagram commutative (here $\overline{d\Phi}$ is $d\Phi(0,0)$ followed by the projection to

$H^1(\Gamma,\mathfrak{g})$).

$$\begin{array}{ccc} \mathbb{R}^r & \xrightarrow{\overline{d\Phi}} & H^1(\Gamma,\mathfrak{g}) \\ \downarrow & \nearrow_{-\delta_*} & \\ \bigoplus_{j=1}^{r} H^0(A_j,\mathfrak{g}) & & \end{array}$$

Proposition 5.1. *Suppose that* ρ *is a stable representation (Section 1) of the fundamental group of a graph of groups into the real points of an algebraic group* $\underline{G}$ *defined over* $\mathbb{R}$ *(with Lie algebra* $\mathfrak{g}$*) such that every edge group has an invariant in* $\mathfrak{g}$ *and no vertex group has a non-zero invariant in* $\mathfrak{g}$. *Then:*

$$\dim X(\Gamma,G) \geq r$$

where r *is the number of edges of the graph*.

Proof. We use the previous lemmas to construct an embedded r-ball B around ρ in $R(\Gamma,G)$. We may assume that B is contained in the set of stable representations. We claim that the image of B in $X(\Gamma,G)$ is the quotient of B by a finite group. To check this, it is sufficient to compute the image of B under the orbit map $\pi: R(\Gamma,G) \to R(\Gamma,G)/G$ since B consists entirely of stable representations. Suppose $\pi(\rho_1) = \pi(\rho_2)$. Then there exists $g \in \underline{G}$ with $\mathrm{Ad}\, g\cdot\rho_1 = \rho_2$. Hence $\mathrm{Ad}\, g\cdot\rho_1|\Gamma_P = \rho_2|\Gamma_P$. Here P is the base vertex (see Lemma 5.7). But by construction $\rho_1|\Gamma_P = \rho_2|\Gamma_P = \rho|\Gamma_P$. Hence $g \in Z(\rho(\Gamma_P))$, the centralizer of $\rho(\Gamma_P)$ in $\underline{G}$. Hence $g \in Z(H)$ where H is the Zariski closure of $\rho(\Gamma_P)$ in $\underline{G}$. But H has no non-zero invariant in $\mathfrak{g}$, hence $Z(H)$ is discrete, hence finite and the proposition is proved since the quotient of an r-ball by a finite group contains a small r-ball.

Remark. If ρ_0 is good then $X(\Gamma,G)$ contains an r-ball around ρ_0.

We are now ready to prove the required lower bounds for the dimension of the spaces of conformal and projective structures on a compact hyperbolic n-manifold M. Let $\Gamma = \pi_1(M)$ and $\rho:\Gamma \to SO(n,1)$ be the standard uniformization. We first treat the case of the space of conformal structures.

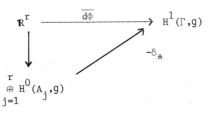

Theorem 5.2. *Suppose* M *contains* r *disjoint, embedded, totally geodesic, two-sided connected hypersurfaces* $M_1, M_2, \ldots, M_r$. *Then the*

dimension of $X(\Gamma, SO(n+1,1))$ *is greater than* r.

Proof. We first check that $\rho(\pi_1(M_j))$ has a non-zero invariant in $so(n+1,1)$ for $j = 1,2,\ldots,r$. We may identify $so(n+1,1)$ with $\Lambda^2\mathbb{R}^{n+2}$ by using $(,)$. Now $\rho(\pi_1(M_j))$ leaves invariant a vector v_j in $\mathbb{R}^{n+1}$ with $(v_j,v_j) > 0$. Also $\rho(\pi_1(M))$ leaves invariant e_{n+1} so a fortiori $\rho(\pi_1(M_j))$ leaves invariant e_{n+1}. Thus $\rho(\pi_1(M_j))$ leaves invariant $v_j \wedge e_{n+1}$. To prove the theorem it suffices to check that $\rho(\pi_1(S_k))$ has no invariant in $so(n+1,1)$ for $k = 1,2,\ldots,b+1$. This follows from the next lemma.

Lemma 5.9. Suppose $n \geq 2$ *and let* S *be a compact hyperbolic manifold with totally geodesic boundary. Let* $\rho:\pi_1(S) \to SO(n,1)$ *be the uniformization representation. Then* $\rho(\pi_1(S))$ *is Zariski dense in* $SO(n,1)$.

Proof. We first prove that $\rho(\pi_1(S))$ has no invariant line in $\mathbb{R}^{n+1}$. Let M be a boundary component of S. Then $\rho(\pi_1(M))$ has a unique invariant line L in $\mathbb{R}^{n+1}$. Since $\pi_1(M) \subset \pi_1(S)$ we see that if $\rho(\pi_1(S))$ has an invariant line then it must be L. Suppose this to be the case. Then $\rho(\pi_1(S))$ is contained in the subgroup H of $SO(n,1)$ which leaves L invariant. Since $\rho(\pi_1(S))$ is discrete in $SO(n,1)$, it is discrete in H. Since $\rho(\pi_1(M))$ is uniform in H so is $\rho(\pi_1(S))$. Hence $M' = \rho(\pi_1(S))\backslash H/K \cap H$ is a compact hyperbolic $(n-1)$-manifold and M is a compact manifold covering M' with $[\pi_1(S):\pi_1(M)]$ sheets. Hence, if we can prove $[\pi_1(S):\pi_1(M)] = \infty$ we are done.

To establish this, assume that $\pi_1(M)$ has finite index in $\pi_1(S)$. The universal cover $\tilde{M}$ of M embeds into the universal cover $\tilde{S}$ of S. Now divide out $\tilde{S}$ by $\rho(\pi_1(M))$. We obtain a cover $S' \to S$ so that the image of $\pi_1(S')$ in $\pi_1(S)$ is precisely $\pi_1(M)$; hence, a finite cover. By construction $M \subset S'$ and the inclusion $\pi_1(M) \to \pi_1(s')$ is an isomorphism. We rename S' by S. We now claim that M is the only boundary component of S. Indeed suppose M' were another. Choose a closed geodesic α' in M'. Because $\pi_1(M)$ maps onto $\pi_1(S)$, α' is freely homotopic to a closed geodesic β in M. Since $M \cap M' = \varphi$, the closed geodesics α' and β are different. But this leads to a contradiction because two different closed geodesics in a hyperbolic manifold are never freely homotopic.

Now we have $M = \partial S$ and the inclusion of $\pi_1(M)$ into $\pi_1(S)$ is an isomorphism. Double S along M to obtain a compact hyperbolic manifold N. By van Kampen's Theorem we have $\pi_1(N) = \pi_1(M)$ but this is impossible because $H_n(\pi_1(M), \mathbb{Z}/2) = 0$ whereas $H_n(\pi_1(N), \mathbb{Z}/2) = \mathbb{Z}/2$.

Now let R be the Zariski closure of $\rho(\pi_1(S))$ in $SO(n,1)$. Then R is not discrete. Also R properly contains H; hence R leaves no totally geodesic subspace of $\mathbb{H}^n$ invariant nor does it fix any point of the closed ball $\mathbb{H}^n \cup S^{n-1}$. Hence, by Theorem 4.4.2 of [7] we have $R \supset SO_0(n,1)$. But R is a real algebraic subgroup of $SO(n,1)$ so $R = SO(n,1)$.

Corollary. *$\rho(\pi_1(S))$ has no non-zero invariant in $so(n+1,1)$.*

Proof. Any invariant of $\rho(\pi_1(S))$ would be an invariant of $R = SO(n,1)$. But $SO(n,1)$ has no invariants in $so(n+1,1)$.

As a consequence of Theorem 5.2 and the Holonomy Theorem, we obtain the following theorems.

Theorem 5.2 (bis). $\dim(\mathbb{C}(M)) \geq r$.

Proof. We have seen that $\mathrm{Hom}(\Gamma,G)/G$ contains embedded r-balls around points arbitrarily close to ρ_0. The theorem now follows from the holonomy theorem.

Theorem 5.2 (tertio). $\dim \mathcal{H}(M \times \mathbb{R}) \geq r$.

Proof. The proof is the same as above.

We now treat the projective case.

Theorem 5.3. *Suppose M contains r disjoint embedded two-sided connected totally geodesic hypersurfaces $M_1, M_2, \ldots, M_r$. Then we have:*

$$\dim X(\Gamma, PGL_{n+1}(\mathbb{R})) \geq r.$$

Proof. We may identify the Lie algebra $\mathfrak{g}$ of $PGL_{n+1}(\mathbb{R})$ with $s\ell_{n+1}(\mathbb{R})$, the Lie algebra of $n+1$ by $n+1$ matrices of trace zero. As a module for $SO(n,1)$, we may identify the $n+1$ by $n+1$ real matrices with $\otimes^2(\mathbb{R}^{n+1})^*$ where the identity matrix is identified with the form $(,)$. Then $so(n,1)$ is identified with $\Lambda^2(\mathbb{R}^{n+1})^*$ and the orthogonal complement $\mathcal{M}$ of $so(n,1)$ in $s\ell_{n+1}(\mathbb{R})$ is identified with $S_0^2(\mathbb{R}^{n+1})^*$, the traceless symmetric 2-tensors. Let $\rho: \Gamma \to SO(n,1) \to PGL_{n+1}(\mathbb{R})$ be the uniformization representation followed by the natural map.

We now observe that $\rho(\pi_1(M_j))$ has a non-zero invariant in $S_0^2(\mathbb{R}^{n+1})^*$ for $j = 1,2,\ldots,r$. We know that in the uniformization representation on $\mathbb{R}^{n+1}$ (or $(\mathbb{R}^{n+1})^*$) the group $\pi_1(M_j)$ has a non-zero invariant v_j. Let h_{v_j} be the traceless projection of the symmetric 2-tensor $v_j \otimes v_j$. Then $\rho(\pi_1(M_j))$ leaves h_{v_j} invariant for $k = 1,2,\ldots,b+1$. But we know $\rho(\pi_1(S_k))$ is Zariski dense in $SO(n,1)$ and $SO(n,1)$ has no non-zero invariant in $S_0^2(\mathbb{R}^{n+1})^*$ - in fact this latter module is irreducible, nor does $SO(n,1)$ have a non-zero invariant in $\Lambda^2(\mathbb{R}^{n+1})^*$. With this the theorem is proved.

As a consequence of Theorem 5.3 and the Holonomy Theorem, we obtain the following theorem.

Theorem 5.3 (bis). $\mathcal{P}(M)$ has dimension greater than or equal to r.

Proof. In the course of the proof of Theorem 5.3 we saw that $\mathrm{Hom}(\Gamma,G)/G$ contained an r-ball around ρ_0. The theorem now follows from the Holonomy Theorem.

6. Singularities in the Deformation Spaces.

In this section, we give a criterion in terms of the topology of M for the spaces $\mathrm{Hom}(\Gamma,G)$ and $\mathrm{Hom}(\Gamma,\underline{G})$ to be singular at a representation ρ and $X(\Gamma,G)$ and $X(\Gamma,\underline{G})$ to be singular at the class of a good representation ρ. In Section 7 we show that this criterion is satisfied for the standard arithmetic examples. In what follows we let $\mathcal{D}$ be a symbol denoting any of the four above spaces.

Lemma 6.1. Suppose M_1 and M_2 are embedded hypersurfaces of M and ρ is any representation of Γ. Suppose the following hold.

(i) $\rho(\pi_1(M_1))$ leaves invariant a non-zero $x \in \mathfrak{g}$.

(ii) $\rho(\pi_1(M_2))$ leaves invariant a non-zero element $y \in \mathfrak{g}$.

(iii) $(M_1 \otimes x)\cdot(M_2 \otimes y) \neq 0$.

Then $\mathrm{Hom}(\Gamma,G)$ and $\mathrm{Hom}(\Gamma,\underline{G})$ are singular at ρ; moreover, if ρ is good then $X(\Gamma,G)$ and $X(\Gamma,\underline{G})$ are singular at the class of ρ.

Proof. For simplicity we assume M_1 is not a boundary and M_2 is not a boundary. Then Γ has an $H\cdot N\cdot N$ decomposition corresponding to M_1 given by $\Gamma = B_1 *_{A_1}$ where $B_1 = \pi_1(M-M_1)$ and $A_1 = \pi_1(M_1)$. Let R_α be the one parameter group in G (or $\underline{G}$) tangent to x. As in Lemma 5.1, we obtain a curve ρ_α in $\mathcal{D}$ constant on B_1 and changing $\rho(\nu_1)$ to its product by R_α. The tangent vector $\dot\rho_\alpha$ to ρ_α at $\alpha = 0$ is, by Theorem 5.1, dual to $M_1 \otimes x$. Let R_β be the one parameter group in G (or $\underline{G}$) tangent to y. Then, as above, we obtain a curve ρ_β in $\mathcal{D}$ leaving B_2 fixed and changing $\rho(\nu_2)$ to its product by R_β. The tangent vector $\dot{\rho}_\beta$ to ρ_β at $\beta = 0$ is dual to $M_2 \otimes y$ by Theorem 5.1. Now consider a linear combination $c\dot{\rho}_\alpha + d\dot{\rho}_\beta$ with $c \neq 0$ and $d \neq 0$. We compute the first obstruction μ (see the end of Section 2) to finding a curve in $\mathcal{D}$ tangent to $c\dot{\rho}_\alpha + d\dot{\rho}_\beta$. We have:

$$\mu = [c\dot{\rho}_\alpha + d\dot{\rho}_\beta, c\dot{\rho}_\alpha + d\dot{\rho}_\beta] = c^2[\dot{\rho}_\alpha,\dot{\rho}_\alpha] + 2cd[\dot{\rho}_\alpha,\dot{\rho}_\beta] + d^2[\dot{\rho}_\beta,\dot{\rho}_\beta].$$

Now $[\dot{\rho}_\alpha,\dot{\rho}_\alpha]$ and $[\dot{\rho}_\beta,\dot{\rho}_\beta]$ are zero because $\dot{\rho}_\alpha$ and $\dot{\rho}_\beta$ are tangent to curves in $\mathcal{D}$. Hence:

$$\mu = 2cd[\dot{\rho}_\alpha,\dot{\rho}_\beta].$$

But by Lemma 4.3, the class $[\dot{\rho}_\alpha,\dot{\rho}_\beta]$ is dual to $(M_1 \otimes x)\cdot(M_2 \otimes y)$. Thus the tangent cone to $\mathcal{D}$ is not a vector space and the lemma is proved.

Remark. In the cases $\mathcal{D} = X(\Gamma,G)$ and $\mathcal{D} = X(\Gamma,\underline{G})$ we must check that the tangent vectors $\dot{\rho}_\alpha$ and $\dot{\rho}_\beta$ are non-trivial and distinct in $H^1(\Gamma,\mathfrak{g})$ (or $H^1(\Gamma,\underline{\mathfrak{g}})$). But this follows because $[\dot{\rho}_\alpha,\dot{\rho}_\beta] \neq 0$ and $[\dot{\rho}_\alpha,\dot{\rho}_\alpha] = 0$.

In this case what is actually proved here is that the slice through ρ in $\mathrm{Hom}(\Gamma,G)$ is not a smooth analytic subvariety of $\mathrm{Hom}(\Gamma_N,G)$ because the tangent cone to the intersection is not a linear subspace of $Z^1(\Gamma,\mathfrak{g})$. This implies that $X(\Gamma,G)$ and $X(\Gamma,\underline{G})$ are singular at $\pi(\rho)$ by the remark following Theorem 1.2.

Before proving the two main theorems of this section we need the following observation. Suppose M_1 and M_3 are disjoint totally geodesic hypersurfaces of M. Let ρ_θ be the deformation of the Fuchsian representation ρ corresponding to the hypersurface M_3. Let v_1 be a non-zero invariant of $\rho(\pi_1(M_1))$. Then v_1 is an invariant of $\rho_\theta(\pi_1(M_1))$ - the curve ρ_θ is constant on $\pi_1(M_1)$ since $M_1 \cap M_3 = \emptyset$. Here we have chosen the base-point of M to lie on M_1. Hence if V_θ denotes the vector space V with Γ acting by ρ_θ then we can form a curve of classes $M_1 \otimes v_1 \in H_{n-1}(M,V_\theta)$. We can now state our main theorems of this section - in what follows we assume the base-point of M is chosen to lie on $M_1 \cap M_2$.

Theorem 6.1. *Suppose M_1,M_2,M_3 are embedded totally geodesic hypersurfaces in M such that $M_1 \cap M_3 = \emptyset$ and $M_2 \cap M_3 = \emptyset$. Let ρ_θ be the deformation of ρ as above corresponding to M_3 and v_1 and v_2*

be non-zero invariants of $\rho(\pi_1(M_1))$ *and* $\rho(\pi_1(M_2))$ *respectively. Assume that for all* θ *the cycle* $M_1 \otimes v_1 \cdot M_2 \otimes v_2$ *is non-zero in* $H_{n-2}(M, \Lambda^2 V_\theta)$.

Then there exists $\varepsilon > 0$ *such that for every* θ *in* $(-\varepsilon, \varepsilon)$ *the point* ρ_θ *(or its class) is a singular point of* $\mathcal{D}$.

Proof. We have only to check that the hypotheses of the previous lemma are satisfied. We take $x = v_1 \wedge e_{n+1}$ and $y = v_2 \wedge e_{n+1}$, then $[x,y] = v_1 \wedge v_2$. The theorem follows since ρ_θ is quasi-Fuchsian but not Fuchsian (hence good) for θ in $(-\varepsilon, \varepsilon) - \{0\}$ for some positive ε.

Corollary. *If* M_1, M_2 *and* M_3 *exist as above then* $\mathcal{D}$ *has non isolated singularities*.

The projective version of Theorem 5.1 goes as follows. We apply Lemma 6.1 with $\underline{G} = PGL_{n+1}$.

Theorem 6.2. *Suppose that for all* θ *the cycle* $M_1 \otimes h_{v_1} \cdot M_2 \otimes h_{v_2}$ *is non-zero in* $H_{n-2}(M, \Lambda^2 V_\theta)$ *(here* h_v *is as in Theorem 5.3)*.

Then there exists $\varepsilon > 0$ *such that for every* θ *in* $(-\varepsilon, \varepsilon)$ *the point* ρ_θ *is a singular point of* $\mathcal{D}$.

Proof. The proof is identical to that of Theorem 6.1.

We conclude this section with a determination of when $[h_v, h_w] = 0$. Recall we are identifying the traceless symemtric 2-tensor h_v with an element of $s\ell(n+1,\mathbb{R})$ using the form $(\,,\,)$. This element is easily seen to be the endomorphism of $\mathbb{R}^{n+1}$ given by:

$$h_v(u) = (u,v)v - \frac{(v,v)}{n+1}u.$$

We find the following formula for the bracket:

$$[h_v, h_w] = (v,w) w \wedge v$$

where by $v \wedge w$ we mean the transformation given by:

$$(w \wedge v) \cdot u = (w,u)v - (v,u)w.$$

Hence $[h_v, h_w] = 0$ if and only if v and w are either proportional or orthogonal. Note that the bracket carries $S_0^2 V$ into $\Lambda^2 V$.

7. Configurations of Totally Geodesic Submanifolds in the Standard Arithmetic Examples.

In this section we verify that the hypotheses of Theorems 6.1 and 6.2 are satisfied for the compact hyperbolic n-manifolds obtained from the standard arithmetic subgroups of $SO(n,1)$. These groups are obtained as follows.

Let p be a positive, square - free integer and $Q:\mathbb{R}^{n+1} \to \mathbb{R}$ be the quadratic form given by:

$$Q(x_1,x_2,\ldots,x_n) = x_1^2 + x_2^2 +\cdots+ x_n^2 - \sqrt{p}\, x_{n+1}^2$$

We let $(,)$ denote the symmetric bilinear form associated to Q. Let $\mathcal{O}$ be the ring of algebraic integers in the quadratic field $k = \mathbb{Q}(\sqrt{p})$. Then the group Φ of matrices with entries in $\mathcal{O}$ which are isometries of Q is a uniform (cocompact) discrete subgroup of the group of matrices with entries in $\mathbb{R}$ which are isometries of Q - see for example Borel [5]. Since this latter group can be identified with $O(n,1)$ in an obvious way, we obtain a uniform, discrete subgroup of $O(n,1)$. The group Φ is often called the group of units of Q, a terminology motivated by the case $n = 1$. By Millson-Raghunathan [15], we can pass to a suitable congruence subgroup $\Gamma = \Gamma(\mathfrak{a})$ of Φ, for $\mathfrak{a}$ an ideal in $\mathcal{O}$, and obtain a uniform, discrete, torsion - free subgroup of $SO_0(n,1)$ and consequently a compact hyperbolic n-manifold $M = \Gamma\backslash\mathbb{H}^n$. We let $\pi:\mathbb{H}^n \to M$ denote the quotient map.

We will use the (upper sheet of the) hyperboloid model for $\mathbb{H}^n$; that is:

$$\mathbb{H}^n = \{z \in \mathbb{R}^{n+1}:(z,z) = -\sqrt{p} \quad \text{and} \quad (z,e_{n+1}) < 0\}.$$

Here $\{e_1,e_2,\ldots,e_{n+1}\}$ is the standard basis of $\mathbb{R}^{n+1}$. We will often write V for $\mathbb{R}^{n+1}$ and L for $\mathcal{O}^{n+1}$, the set of vectors with coordinates in $\mathcal{O}$.

We now construct compact orientable non-singular totally geodesic submanifolds in suitable (congruence subgroup) covers of M. Let $X = \{x_1,x_2,\ldots,x_k\}$ be a k-tuple of vectors in L chosen so that:

(i) $\dim \operatorname{span} X = k$

(ii) $(,)|\operatorname{span} X$ is positive definite.

We let $\mathbb{H}_X^n$ denote the totally geodesic, codimension k submanifolds of $\mathbb{H}^n$ given by:

$$\mathbb{H}_X^n = \{z \in \mathbb{H}^n:(z,x) = 0 \quad \text{for} \quad \operatorname{span} X\}$$

We put $M_X = \pi(\mathbb{H}^n_X)$. Usually M_X will have self-intersections; however the following lemma shows that the self-intersections may be removed upon passing to a suitable cover. In what follows we let r_X denote the involution of V given by:

$$r_X(x) = -x \quad \text{for} \quad x \in \text{span } X$$
$$r_X(x) = x \quad \text{for} \quad x \in (\text{span } X)^\perp.$$

Then $\mathbb{H}^n_X$ is the fixed-point set of r_X acting on $\mathbb{H}^n$. A subscript X on a subgroup of $SO(n,1)$ will denote the subgroup of elements fixing $\{x_1,x_2,\ldots,x_k\}$. In particular $\Gamma_X = \{\gamma \in \Gamma : \gamma x_j = x_j$ for $j = 1,2,\ldots,k\}$. For a subgroup $\Gamma' \subset \Gamma$ we let $\pi' : \mathbb{H}^n \to M' = \Gamma' \backslash \mathbb{H}^n$ denote the quotient map. A prime superscript on an object in M which is the image under π of an object on $\mathbb{H}^n$ will denote the corresponding image under π'; for example, $M'_X = \pi'(\mathbb{H}^n_X)$ and $M'_Y = \pi'(\mathbb{H}^n_Y)$.

Lemma 7.1. *There exists a congruence subgroup* $\Gamma' \subset \Gamma$ *so that* $\pi'(\mathbb{H}^n_X) = \Gamma'_X \backslash \mathbb{H}^n$. *In this case* $\pi'(\mathbb{H}^n_X)$ *is an orientable submanifold. Moreover if* $\Gamma'' \subset \Gamma'$ *and* $\gamma'' \in \Gamma''$ *satisfies* $\gamma''\mathbb{H}^n_X \cap \mathbb{H}^n_X \neq \emptyset$ *then* $\gamma'' \in \Gamma''_X$.

Proof. Choose Γ' so that $r_X \Gamma' r_X = \Gamma'$. By the Jaffee Lemma, Millson [14], Lemma 2.1, we find that $\pi'(\mathbb{H}^n_X) = \Lambda \backslash \mathbb{H}^n_X$ where $\Lambda = \{\gamma \in \Gamma' : r_X \gamma r_X = \gamma\}$; that is, γ preserves the splitting $V = \text{span } X + (\text{span } X)^\perp$. But consider the action on span X induced by Λ. The projection of Λ is a discrete subgroup of the direct product of the orthogonal group of span X with itself (because the projection of Λ leaves invariant a lattice in span $X \oplus$ span X). But the restriction of Q to span X is positive definite and consequently the projection of Λ is finite. Hence if Γ is neat (so no element of Γ has an eigenvalue equal to a non-trivial root of unity) we find that $\Lambda = \Gamma'_X$. With this the first statement is proved. The second statement follows because Γ'_X preserves the orientation of $\mathbb{H}^n_X$ (see remarks below) and Γ'_X is torsion free. The third statement follows because $\Gamma'' \cap \Gamma'_X = \Gamma''_X$.

Remark. In the course of the proof, we showed that if γ preserves span X and Γ is neat then γ fixes the elements of X.

To orient M_X it is sufficient to orient $\mathbb{H}^n_X$. The normal bundle of $\mathbb{H}^n_X$ may be canonically identified with span X; thus, it is sufficient to orient span X. We orient $\mathbb{H}^n_X$ so that the orientation of $\mathbb{H}^n_X$ at z followed by the orientation of span X followed by z is the orientation of the standard basis of V.

We now rename Γ' by Γ and suppress all primes. By Millson [14], Section 4, for any positive integer m we can find a cover of M containing at least m disjoint non-singular orientable totally geodesic hypersurfaces (which in addition are homologically independent). By the results of Section 5, we deduce the following theorem.

Theorem 7.1. *For any* $m > 0$ *and any* $n \geq 2$, *there exists a compact hyperbolic* n-*manifold* M *with fundamental group* Γ, *a standard arithmetic subgroup of* $SO(n,1)$, *such that the dimensions of* $\mathcal{C}(M)$, $\mathcal{P}(M)$, $\mathcal{H}(M \times R)$, $X(\Gamma, SO(n+1,1))$ *and* $X(\Gamma, PGL_{n+1}(R))$ *are all greater than or equal to* m.

We now assume $X = \{x_1,\ldots,x_p\}$ and $Y = \{y_1,\ldots,y_q\}$ are chosen so that $X \cup Y$ spans a subspace U of dimension $p+q$ so that $(,)|U$ is positive definite. This assumption on U implies that $\mathbb{H}^n_X$ and $\mathbb{H}^n_Y$ intersect transversely in a codimension $p+q$ totally geodesic subspace. We do not assume that the vectors in X are orthogonal to those in Y. We let E and F be flat bundles over M and $v: E \otimes E \to F$ be a parallel bundle map as in Section 4. We choose a point w_0 on $\mathbb{H}^n_X \cap \mathbb{H}^n_Y$ as a base-point for $\mathbb{H}^n$ and let $z_0 = \pi(w_0)$ be a base-point for M. We let E_0 and F_0 denote the fibers of E and F over z_0. We assume Γ_X has an invariant α_X in E_0 and Γ_Y has an invariant β_Y in F_0. The invariant α_X corresponds to a parallel section s_X of $E|M_X$ satisfying $s_X(z_0) = \alpha_X$. The invariant β_Y corresponds to a parallel section s_Y of $E|M_Y$ satisfying $s_Y(z_0) = \beta_Y$. Then $v(s_X, s_Y)$ is a parallel section of $F|M_X \cap M_Y$. We now give a formula for the intersection cycle $(M_X \otimes s_X)\cdot(M_Y \otimes s_Y)$. We assume M_X and M_Y intersect transversely in disjoint codimension $p+q$ submanifolds $P_1, P_2, \ldots, P_\ell$. We first show how to orient each P_j.

Choose an orientation ω for P_j. The orientation ω induces an orientation ω_1 of the normal bundle of P_j in M_X and an orientation ω_2 of the normal bundle of P_j in M_Y by requiring that ω followed by ω_1 be the orientation of M_X and ω followed by ω_2 be the orientation of M_X. Then $\omega_1 \wedge \omega_2$ is independent of the choice of ω. We define $\varepsilon(\omega)$ to be $+1$ if the orientation of $\omega \wedge \omega_1 \wedge \omega_2$ is the orientation of M and $\varepsilon(\omega)$ to be -1 otherwise. We will call the orientation of P_j such that $\varepsilon(\omega) = +1$ the intersection orientation.

Remark. The intersection orientation may also be described as the orientation ω for P_j so that the induced orientation of the normal bundle of P_j in M_X coincides with that of the restriction to P_j of the normal bundle of M_Y in M - note that this second bundle already has an orientation.

We give each component P_j for $j = 1,2,\ldots,\ell$, the intersection orientation. We then have an equality of oriented cycles:

$$M_X \cdot M_Y = \sum_{j=1}^{\ell} P_j .$$

By definition of the intersection of cycles with coefficients we also have:

$$(M_X \otimes s_X)\cdot(M_Y \otimes s_Y) = \sum_{j=1}^{\ell} P_j \otimes v(s_X,s_Y)|P_j$$

We wish to obtain a formula which will enable us to determine when $v(s_X,s_Y)|P_j$ is zero.

In order to simplify notation we suppress the subscript j, replacing P_j by P and γ_j by γ. We let t denote the section $v(s_X,s_Y)$ of $F|P$. We have chosen a component $\tilde{P} = \gamma(H_X^n) \cap H_Y^n$ of $\pi^{-1}(P)$. We lift t to a section $\tilde{t}$ of the pull-back of F to $\tilde{P}$. We then parallel translate $\tilde{t}$ to w_0 and evaluate, obtaining an element $\varphi(t|P)$ which is zero if and only if $v(s_X,s_Y)|P$ is zero. We wish to evaluate $\varphi(t|P)$ in terms of γ, α_X and β_Y. Choose $w_2 \in \tilde{P}$, let $z_2 = \pi(w_2)$ and let $w_1 = \gamma^{-1}(w_2)$ so $w_1 \in H_X^n$. We choose a path $\tilde{a}$ in H_X^n from w_0 to w_1 and a path $\tilde{b}$ in H_Y^n from w_0 to w_2. We let $a = \pi(\tilde{a})$ and $b = \pi(\tilde{b})$. Then ab^{-1} represents γ^{-1} in $\pi_1(M,z_0)$ since it lifts to $\tilde{a}\gamma^{-1}(\tilde{b}^{-1})$. By definition $s_X(z_2) = a_* X$ where a_* denotes parallel translation along a. Also $s_Y(z_2) = b_*\beta_Y$. Hence $t(z_2) = v(a_*\alpha_X,b_*\beta_Y)$ and hence $\tilde{t}(w_2) = v(a_*\alpha_X,b_*\beta_Y)$. We obtain $\varphi(t|P)$ by parallel translating $\tilde{t}(w_2)$ back along $\tilde{b}$; that is $\varphi(t|P) = (\tilde{b}^{-1})_* v(a_*\alpha_X,b_*\beta_Y)$. But $(\tilde{b}^{-1})_* = (b^{-1})_*$ and hence $\varphi(t|P) = (b)_*^{-1} v(a_*\alpha_X,b_*\beta_Y) = v(b_*^{-1}a_*\alpha_X,\beta_Y) = v((ab^{-1})_*\alpha_X,\beta_Y)$. Now $(ab^{-1})_*\alpha_X$ is the parallel translate of α_X around a loop representing γ^{-1}. This is the way γ acts on α_X via its action on the standard fiber. We obtain the following lemma.

Lemma 7.2. $v(s_X,s_Y)|P = 0$ *if and only if* $v(\gamma\alpha_X,\beta_Y) = 0$ *in* E_0.

Remark. If we choose a different γ, say $\gamma' = \eta\gamma\mu$ with $\eta \in \Gamma_Y$ and $\mu \in \Gamma_X$, then P would change to ηP and the coefficient would change to $\eta v(\gamma\alpha_X, \beta_Y)$.

We define a subset $\Delta \subset \Gamma$ by:

$$\Delta = \{\gamma \in \Gamma : \gamma(\mathbb{H}_X^n) \cap \mathbb{H}_Y^n \neq \emptyset\}.$$

Then $\Gamma_X \times \Gamma_Y$ acts on Δ by $(\gamma_1, \gamma_2) \cdot \gamma = \gamma_2 \gamma \gamma_1^{-1}$. The map $\gamma \to \pi(\gamma(\mathbb{H}_X^n) \cap \mathbb{H}_X^n)$ induces a one-to-one correspondence between the orbits of $\Gamma_X \times \Gamma_Y$ in Δ and the components of $M_X \cap M_Y$. Hence Δ consists of a finite number of $\Gamma_X \times \Gamma_Y$ orbits (or Γ_Y, Γ_X double cosets). For any ideal $\mathfrak{b} \subset \mathcal{O}$ we define:

$$\Delta(\mathfrak{b}) = \Delta \cap \Gamma(\mathfrak{b}).$$

We observe that if $\mathfrak{c} \subset \mathfrak{b}$ then $\Delta(\mathfrak{c}) \subset \Delta(\mathfrak{b})$.

We have the following theorem under the assumptions $p+q \neq n-1$ and $n \geq 4$. In the next theorem we consider congruence subgroups $\Gamma'' \subset \Gamma' \subset \Gamma$. We let $M' = \Gamma' \backslash \mathbb{H}^n$ and $M'' = \Gamma'' \backslash \mathbb{H}^n$. We let π' and π'' be the covering projections and $M'_X = \pi'(\mathbb{H}_X^n)$ and $M''_X = \pi''(\mathbb{H}_X^n)$ and similarly for Y. We assume in what follows that Γ is the congruence subgroup of Φ, the group of units of $(,)$, of level $\mathfrak{a}$.

<u>Theorem 7.2</u>. <u>There exists a congruence cover</u> M' <u>of</u> M <u>so that</u> $M'_X \cap M'_Y$ <u>consists of the single component</u> $\pi'(\mathbb{H}_X^n \cap \mathbb{H}_Y^n)$. <u>Moreover for any congruence cover</u> M'' <u>of</u> M' <u>the intersection</u> $M''_X \cap M''_Y$ <u>again consists of the single component</u> $\pi''(\mathbb{H}_X^n \cap \mathbb{H}_Y^n)$.

Theorem 7.2 will be a consequence of the following proposition.

<u>Proposition 7.1</u>. <u>If</u> $p+q \neq n-1$, <u>there exists an ideal</u> $\mathfrak{b}$ <u>so that</u> $\Delta(\mathfrak{b}) \subset \Gamma_Y \Gamma_X$.

In what follows $R_{\mathfrak{b}}(.)$ will denote reduction modulo the ideal $\mathfrak{b}$. We define $\Delta' \subset L^p$ by:

$$\Delta' = \{X' \in L^p : X' = \gamma' X \text{ for some } \gamma' \in \Delta\}.$$

The proof of Proposition 7.1 will follow two lemmas. The next lemma shows how to pass to a congruence cover and eliminate certain intersection components.

<u>Lemma 7.3</u>. <u>Let</u> $X' \in \Delta'$ <u>and suppose</u> $\mathfrak{b} \subset \mathcal{O}$ <u>is an ideal such that</u>:

$$R_{\mathfrak{b}}(\Gamma_Y X') \cap R_{\mathfrak{b}}(\Gamma_Y X) = \emptyset.$$

<u>Then</u> $\Delta(\mathfrak{b}) \cap \Gamma_Y \gamma' \Gamma_X = \emptyset$ <u>where</u> $\gamma' \in \Delta$ <u>satisfies</u> $\gamma' X = X'$.

Proof. If $\gamma \in \Delta(\mathfrak{b}) \cap \Gamma_Y\gamma'\Gamma_X$ so $\gamma = \gamma_2\gamma'\gamma_1$ with $\gamma_2 \in \Gamma_Y$ and $\gamma_1 \in \Gamma_X$ then:

$$R_{\mathfrak{b}}(X) = R_{\mathfrak{b}}(\gamma X) = R_{\mathfrak{b}}(\gamma_2\gamma'\gamma_1 X) = R_{\mathfrak{b}}(\gamma_2 X').$$

With this the lemma is proved.

We now use Lemma 7.3 to eliminate all double cosets so that the orbit of X under the double coset can be separated modulo some ideal $\mathfrak{c}$ from the trivial double coset.

Lemma 7.4. There exists an ideal $\mathfrak{b}$ so that $\gamma \in \Delta(\mathfrak{b})$ implies $R_{\mathfrak{c}}(\Gamma_Y\gamma X) \cap R_{\mathfrak{c}}(\Gamma_Y X) \neq \emptyset$ for any $\mathfrak{c}$.

Proof. There are a finite number of Γ_Y, Γ_X double cosets in Δ. Let $\{\gamma_1,\gamma_2,\ldots,\gamma_r\}$ be a set of representatives. Either there exists an ideal $\mathfrak{b}_1$ such that $R_{\mathfrak{b}_1}(\Gamma_Y\gamma_1 X) \cap R_{\mathfrak{b}_1}(\Gamma_Y X) = \emptyset$ or no such ideal exists. If such an ideal exists then by Lemma 7.3 we know $\Delta(\mathfrak{b}_1) \cap \Gamma_Y\gamma_1\Gamma_X = \emptyset$ and we have eliminated the double coset containing γ_1. If no such ideal exists then for every element $\gamma \in \Gamma_Y\gamma_1\Gamma_X$ we have $R_{\mathfrak{c}}(\Gamma_Y\gamma X) \cap R_{\mathfrak{c}}(\Gamma_Y X) \neq \emptyset$ for every $\mathfrak{c}$. In this case we do not need to eliminate γ_1 so we take $\mathfrak{b}_1 = \mathcal{O}$. Continuing in this way we obtain ℓ (possibly non-proper) ideals $\mathfrak{b}_1,\mathfrak{b}_2,\ldots,\mathfrak{b}_\ell$. We put $\mathfrak{b} = \mathfrak{b}_1\mathfrak{b}_2\ldots\mathfrak{b}_\ell$ and the lemma is proved.

We now begin the proof of Proposition 7.1. For the course of this proof Γ_X and Γ_Y will be denoted Γ_1 and Γ_2 and G_X and G_Y by G_1 and G_2. For a ring R containing $\mathcal{O}$, the symbol $G_1(R)$ will denote the R-rational points of the algebraic subgroup of $SO(Q)$ fixing X and similarly for $G_2(R)$. The symbol $\mathcal{O}_P$ will denote the P-adic completion of $\mathcal{O}$ and the symbol $G_1(\mathcal{O}_P,\mathfrak{a})$ will denote the subgroup of $G_1(\mathcal{O}_P)$ consisting of γ satisfying $\gamma \equiv 1 \bmod P^m$ where m is the largest power of P dividing $\mathfrak{a}$. We will assume Γ is chosen so that $\gamma \in \Gamma$ implies that the k spinor norm of γ is 1; this is possible by Millson-Raghunathan [16], Proposition 4.1.

The idea of the proof is to examine the elements $\gamma \in \Delta(\mathfrak{b})$; that is, elements γ such that the corresponding vectors $X' = \gamma' X$ have an associated Γ_2 orbit, $\Gamma_2 X'$, which cannot be separated modulo any ideal $\mathfrak{c}$ from $\Gamma_2 X$. We show all such γ' satisfy $\gamma' \in \Gamma_2\Gamma_1$. We let

$$\Delta'(\mathfrak{b}) = \{X' : X' = \gamma X \text{ for } \gamma \in \Delta(\mathfrak{b})\}.$$

If $X \in V^p$, $Y \in V^q$, the symbol (X,Y) denotes the matrix

$((x_i,y_j))$. For $X' \in \Delta'$, let $B(X') = (X',Y)$, a p by q matrix with entries in k. Then for $\gamma_2 \in \Gamma_2$ we have $B(\gamma_2 X') = B(X')$ and B is constant on Γ_2 orbits in Δ'. If $B(X') \neq B(X)$ then there exists some $y_j \in Y$ and some x_i' in X with $(x_i',y_j) \neq (x_i,y_j)$. Hence, for almost every prime P in $\mathcal{O}$ we have $(x_i',y_j) \not\equiv (x_i,y_j) \bmod P$ and consequently $R_P(\gamma_2 X') \neq R_P(X)$ for $\gamma_2 \in \Gamma_2$. Hence $R_P(\Gamma_2 X')$ $R(\Gamma_2 X) = \emptyset$. Thus, if $X' \in \Delta'(b)$, we have $B(X') = B(X)$.

But if $X' \in \Delta'$ then $X' = \gamma X$ with $\gamma \in \Delta$ so $(X',X') = (X,X)$. Consequently, the matrix of inner products of $(,)$ relative $X' \cup Y$ is the same as the matrix of $(,)$ relative $X \cup Y$. Consequently, if $X' \in \Delta'(b)$, there exists $g \in G(k)$ such that $gX = X'$ and $gY = Y$.

We claim, that in case $p + q \leq n - 2$, we may assume that g has spinor norm 1. For, in this case, the orthogonal complement W of span$(X \cup Y)$ is an indefinite space of dimension greater than or equal to 3. Hence, by O'Meara [18], 101·8, we may find an element $\eta \in SO(W)$ with entries in k and having the same spinor norm as g. Then, replacing g by $g\eta$, we prove the claim (we will need this later in the case $p + q \leq n - 2$).

In any case, since $gY = Y$, we have $g = g_2 \in G_2(k)$. But then $g_2^{-1}\gamma' X = X$ so $g_2^{-1}\gamma' = g_1 \in G_1(k)$ and we obtain:

$$\gamma' = g_2 g_1 \in G_2(k)G_1(k).$$

By definition, if $X' \in \Delta'(b)$, we may suppose that for every prime ideal P in $\mathcal{O}$ and every integer $m > 0$ there exists an element $\gamma_2 = \gamma_2(P,m)$ with:

$$R_{P^m}(\gamma_2 X') = R_{P^m}(X).$$

The infinite set $\{\gamma_2(P,m)\} \subset \Gamma_2 \subset G_2(\mathcal{O}_P,a)$ has a limit point ν_P^{-1} in $G_2(\mathcal{O}_P,a)$ satisfying $\nu_P X = X'$. We may assume that the spinor norm of ν_P is 1 since the kernel of the spinor norm is closed in $G_2(\mathcal{O}_P)$ - it is the intersection of $G_2(\mathcal{O}_P)$ with the image of the spin group in $G_2(\mathcal{O}_P)$. But then defining $\mu_P = \nu_P^{-1}\gamma$ we find that $\mu_P \in G_1(\mathcal{O}_P,a)$ and:

$$\gamma' = \nu_P \mu_P.$$

At this point we separate the proof of the theorem into two cases; the first in which $p + q = n$ and the second in which $p + q \leq n - 2$.

For the first case we note $G_1 \cap G_2 = \{1\}$ since any $g \in G_1 \cap G_2$ fixes a subspace of codimension 1 and has determinant 1. Thus, we must have:

$$g_2 = \nu_P$$

and

$$g_1 = \mu_P$$

for all P. This concludes the proof of the theorem for the case $p + q = n$ since the above equality implies $g_1 \in \Gamma_1$ and $g_2 \in \Gamma_2$.

In case $p + q < n - 1$ we consider the adele $\{a_P\}$ with P th component a_P given by:

$$a_P = \nu_P^{-1} g_2.$$

Then a_P has spinor norm 1 by the previous claim and we may apply the Strong Approximation Theorem to the algebraic group $H = SO(W)$, see O'Meara [19], to conclude that there exists $\eta \in H(k)$ and an adele $\{b_P\} \in \Pi\, H(\mathcal{O}_P, a)$ such that:

$$\{b_P \eta^{-1}\} = \{\nu_P^{-1} g_2\}$$

From the previous equation we deduce $\nu_P b_P = g_2\eta$. Consequently $g_2\eta$ is an element of Γ fixing Y and so $g_2\eta \in \Gamma_2$. Renaming $g_2\eta$ by ν and defining $\mu = \nu^{-1}\gamma$ we find $\mu \in \Gamma_1$ and $\gamma = \nu\mu \in \Gamma_2\Gamma_1$. With this Proposition 7.1 is proved.

We now show how Proposition 7.1 implies Theorem 7.2. Choose b so that $\Delta(b) \subset \Gamma_Y\Gamma_X$. Suppose first $p + q = n$. Then $\nu\mu X = \nu X$ and $\gamma X = \nu\mu X \equiv X \bmod b$. Hence $\nu X \equiv X \bmod b$. But also $\nu Y = Y$. Hence $\nu \equiv 1 \bmod b$ on span$(X \cup Y)$. But this span has codimension 1 and $\det \nu = 1$. Hence $\nu \equiv 1 \bmod b$ and consequently $\mu = \nu^{-1}\gamma$ also satisfies $\mu \equiv 1 \bmod b$.

Suppose now that $p + q \leq n - 2$. The previous argument shows that ν and μ are congruent to 1 modulo b on span$(X \cup Y)$. Also $\nu \equiv \mu^{-1} \bmod b$ on W. Let φ be the element of the finite group of isometries of W modulo b to which ν and μ^{-1} are congruent. Since φ has spinor norm 1 and the dimension of W is greater than or equal to 3, by the Strong Approximation Theorem we may find $\eta \in \Gamma \cap SO(W)$ so that $\eta^{-1} \equiv \varphi \bmod b$. We let $\nu' = \nu\eta$ and $\mu' = \eta^{-1}\mu$. Then $\nu' \equiv \mu' \equiv 1 \bmod b$ and $\gamma = \nu'\mu'$. This proves the first part of Theorem 7.2.

To prove the second part note that if $c \subset b$ then $\Delta(c) \subset \Delta(b) \subset \Gamma_Y\Gamma_X$ and we may repeat the previous argument.

We now apply Theorem 7.2 to the case $X = \{e_1, e_2\}$ and

$Y = \{y_1, y_2, \ldots, y_{n-2}\}$ an $\mathcal{O}$-integral $(n-2)$-frame chosen so that $X \cup Y$ spans a positive definite space of dimension n and so that $(e_1 \wedge e_2, y_1 \wedge y_2) > 0$. For example take $Y = \{e_1 + e_3, e_2 + e_4, e_5, \ldots, e_n\}$. We may assume, by the remark following Lemma 7.1, that Γ_Y acts trivially on the span of Y. Consequently we may form a cycle with coefficients in $\Lambda^2 V$ given by $M_Y \otimes y_1 \wedge y_2$. Similarly we have a cycle with coefficients in $\Lambda^2 V$ given by $M_Y \otimes e_1 \wedge e_2$. We use the form induced by $(,)$ on $\Lambda^2 V$ to define

$$(M_X \otimes e_2 \wedge e_2) \cdot (M_Y \otimes y_1 \wedge y_2).$$

Lemma 7.5. *There exists a congruence subgroup $\Gamma(b) \subset \Gamma$ so that the corresponding cycles $M_X' \otimes e_1 \wedge e_2$ and $M_Y' \otimes y_1 \wedge y$ satisfy:*

$$(M_X' \otimes e_1 \wedge e_2) \cdot (M_Y' \otimes y_1 \wedge y_2) \neq 0.$$

Proof. By Theorem 7.2 we may find b so that M_X' and M_Y' intersect at $\pi'(\mathbb{H}_X^n \cap \mathbb{H}_Y^n)$. We apply Lemma 7.2 with $\gamma = 1$ and find the coefficient contribution $(e_1 \wedge e_2, y_1 \wedge y_2) \neq 0$. With this Lemma 7.5 is proved.

Corollary. $M_X' \otimes e_1 \wedge e_1$ *is a non-zero class in* $H_{n-2}(\Gamma(b)\, \mathbb{H}^n, \Lambda^2 V)$.

We replace our original Γ by $\Gamma(b)$ and suppress all primes. We now apply Theorem 7.2 to the case $X = \{e_1\}$ and $Y = \{e_2\}$. We consider the cycles with coefficients in V given by $M_{e_1} \otimes e_1$ and $M_{e_2} \otimes e_2$. We use the exterior product from $V \otimes V$ to $\Lambda^2 V$ to define $(M_{e_1} \otimes e_1) \cdot (M_{e_2} \otimes e_2)$ as an element of $H_{n-2}(M, \Lambda^2 V)$. Let us denote $\pi(\mathbb{H}^n_{\{e_1, e_2\}}) \otimes e_1 \wedge e_2$ by $Z \otimes e_1 \wedge e_2$.

Remark. $Z \otimes e_1 \wedge e_2$ is not a boundary; hence if $\Gamma' \subset \Gamma$ is a subgroup of finite index, $\pi' : \mathbb{H}^n \to \Gamma' \backslash \mathbb{H}^n$ is the covering and $Z' = \pi'(\mathbb{H}^n_{e_1} \cap \mathbb{H}^n_{e_2})$ then $Z' \otimes e_1 \wedge e_2$ is not a boundary.

Lemma 7.6. *There exists a congruence subgroup $\Gamma(c) \subset \Gamma$ so that:*

$$(M_{e_1}' \otimes e_1) \cdot (M_{e_2}' \otimes e_2) \neq 0.$$

Proof. We apply Theorem 7.2 to deduce that there exists a congruence subgroup $\Gamma(c) \subset \Gamma$ so that $M_{e_1}' \cap M_{e_2}' = \pi'(\mathbb{H}^n_{e_1} \cap \mathbb{H}^n_{e_2})$. We denote this intersection by Z'. By the previous remark $Z' \otimes e_1 \wedge e_2 \neq 0$ and the lemma is proved since by Lemma 7.2 the coefficient contribution is non-zero - again applying Lemma 7.2 with $\gamma = 1$.

We have now proved the desired non-vanishing theorem for intersection products of hypersurfaces with coefficients in V. As a consequence of the results of Section 6 we have the following theorem, again assuming $n \geq 4$.

Theorem 7.3. $\mathrm{Hom}(\Gamma, SO(n+1,1))$ and $\mathrm{Hom}(\Gamma, \underline{SO}(n+1,1))$ each have a singularity at ρ_0.

Similar arguments based on Theorem 7.1 using coefficients in $S_0^2(V)$ yield the required theorem for projective structures. Note that $[h_{e_1}, h_{e_1+e_2}] = e_1 \wedge e_2$.

Lemma 7.7. For any subgroup Γ of finite index in the units of (,) *there exists a further congruence subgroup $\Gamma(c)$ so that:*

$$(M'_{e_1} \otimes h_{e_1}) \cdot (M'_{e_1+e_2} \otimes h_{e_1+e_2}) \neq 0.$$

Remark. In fact we obtain $(M'_{e_1} \otimes h_{e_1}) \cdot (M'_{e_1+e_2} \otimes h_{e_1+e_2}) = M_{\{e_1,e_2\}} \otimes e_1 \wedge e_2$ which we proved to be non-zero in Lemma 7.5.

Since ρ_0 is good in the projective case, we obtain the following theorem, again assuming $n \geq 4$.

Theorem 7.4. $\mathrm{Hom}(\Gamma, PGL_{n+1}(\mathbb{R}))$, $\mathrm{Hom}(\Gamma, PGL_{n+1}(\mathbb{C}))$, $X(\Gamma, PGL_{n+1}(\mathbb{R}))$ and $X(\Gamma, PGL_{n+1}(\mathbb{C}))$ are singular at ρ_0.

We now wish to establish the existence of non-isolated singularities for the deformation spaces. By the results of Section 6, it is sufficient to find a two-sided, totally geodesic, non-singular hypersurface N disjoint from M_{e_1} and M_{e_2} (or M_{e_1} and $M_{e_1+e_2}$). We prove a more general theorem in the framework of Theorem 7.2 with X and Y as in that theorem. We suppose that $f: M' \to M$ is a cover and that Ψ is the group of covering transformations of f. We let Ψ_X denote the group of covering transformations of $M'_X \to M_X$ and Ψ_Y denote the group of covering transformations of $M'_Y \to M_Y$. Let $\eta \in \Psi$.

Lemma 7.8. $\eta(M'_X) \cap M'_X \neq \emptyset$ if and only if $\eta \in \Psi_X$.

Proof. Suppose $\eta(M'_X) \cap M'_X \neq \emptyset$. We choose $\gamma \in \Gamma$ representing η. Then there exist x_1 and x_2 in $\mathbb{H}^n_X$ such that $\pi'(\gamma x_1) = \pi'(x_2)$. Hence there exists $\gamma' \in \Gamma'$ so that $\gamma'\gamma x_1 = x_2$. But by Lemma 7.1, we have $\gamma'\gamma \in \Gamma_X$ and consequently $\eta \in \Psi_X$. The converse is clear and the lemma is proved.

We now examine when $\eta(M'_X)$ meets M'_Y. We assume M is chosen to satisfy the conclusions of Theorem 7.2; that is, we require that

$M_X \cap M_Y$ consist of a single component.

Lemma 7.9. $\eta(M'_X) \cap M'_Y \neq \emptyset$ if and only if $\eta \in \Psi_Y \Psi_X$.

Proof. Suppose $\eta(M'_X) \cap M'_Y \neq \emptyset$. Choose $\gamma \in \Gamma$ representing η. Then there exist $x_1 \in H^n_X$ and $x_2 \in H^n_Y$ such that $\pi'(x_1) = \pi'(x_2)$. Hence there exists $\gamma' \in \Gamma'$ such that $\gamma'\gamma x_1 = x_2$; that is, $\gamma'\gamma \in \Delta$. But by construction of Γ (Theorem 7.2) there exist $\nu \in \Gamma_Y$ and $\mu \in \Gamma_X$ such that $\gamma'\gamma = \nu\mu$. Reducing modulo Γ' we find $\eta \in \Psi_Y \Psi_X$. The converse is clear and the lemma is proved.

Theorem 7.5. Let $Y_1, Y_2, \ldots, Y_m$ be given such that for $j = 1,2,\ldots,m$ the set $X \cup Y_j$ spans a positive definite subspace of V of dimension $p+q$ with $p + q \neq n - 1$. Then there exists a covering $f: M' \to M$ and a covering transformation η of f such that $\eta(M'_X)$ does not intersect $M'_X, M'_{Y_1}, \ldots, M'_{Y_m}$.

Proof. We apply Theorem 7.2 successively to $Y_1, \ldots, Y_m$ to arrange that $M'_X \cap M'_{Y_j}$ consists of a single component for $j = 1,2,\ldots,m$. From Lemma 7.9 we find that it is sufficient to find a covering group Ψ such that $\Psi \neq \Psi_{Y_1}\Psi_X \cup \cdots \cup \Psi_{Y_n}\Psi_X$. Suppose no such cover exists. Choose $x \in X$ and $y_j \in Y_j$ for $j = 1,2,\ldots,m$. Then the equation $\Pi^m_{j=1}[(gx,y_j) - (x,y_j)] = 0$ is satisfied for all g in the congruence completion of Γ, hence for all $g \in \Gamma$ and hence by Zariski density for all $g \in \underline{G}$. Since $\underline{G}$ is irreducible one of the factors in the above equation must vanish identically on $\underline{G}$. But this is absurd.

We now prove the main theorem of this section assuming $n \geq 4$ and Γ as above.

Theorem 7.6. The spaces $R(\Gamma, SO(n+1,1))$, $R(\Gamma, \underline{SO}(n+1,1))$, $X(\Gamma, SO(n+1,1))$, $X(\Gamma, \underline{SO}(n+1,1))$, $R(\Gamma, PGL_{n+1}(\mathbb{R}))$, $R(\Gamma, PGL_{n+1}(\mathbb{C}))$, $X(\Gamma, PGL_{n+1}(\mathbb{R}))$ and $X(\Gamma, PGL_{n+1}(\mathbb{C}))$ all have non-isolated singularities.

Proof. We give the proof for the first case. We apply the previous theorem to the case $X = \{e_1\}$, $Y = \{e_2\}$ and $Y_2 = \{e_2 + e_4\}$. Then $\eta(M'_{e_1})$ is a totally geodesic hypersurface which does not intersect M'_{e_1}, M'_{e_2} or $M'_{e_2+e_4}$ and a fortiori does not intersect the surface $M'_{\{e_1+e_3, e_2+e_4, e_5, e_6, \ldots, e_m\}}$. Hence if ρ_t is the deformation of ρ corresponding to the totally geodesic hypersurface $\eta(M'_{e_1})$ then ρ_t is constant on the fundamental groups of the three above manifolds and the intersection number calculations of Lemma 7.5 and Lemma 7.6 are independent of t.

Remark. Further work is required in order to make precise the statement that $\mathcal{C}(M)$, $\mathcal{H}(M \times \mathbb{R})$ and $\mathcal{P}(M)$ are singular. First, we need a "completeness" theorem to the effect that each point in $\mathcal{S}(M)$, the space of marked (G,X) structures for some G and X, has a neighborhood isomorphic to an analytic subvariety in $H^1(M,\Theta)$ where Θ is the sheaf of infinitesimal automorphisms of the G-structure. Second we need to know that the holonomy map preserves this structure. It appears that these results can be proved by imitating the proof of completeness for complex structures.

8. $\mathcal{C}(M)$ and Riemannian Geometry.

In this section, we will regard $\mathcal{C}(M)$ as the quotient space of Riemannian metrics with vanishing Weyl tensor by the group which is the semi-direct product of the group $C_+^\infty(M)$ of strictly positive smooth functions on M and the group of diffeomorphisms of M isotopic to the identity. Thus a point $c \in \mathcal{C}(M)$ is an equivalence class of Riemannian metrics all of which have zero Weyl tensor. In what follows n will denote the dimension of the manifold M under consideration. The following theorem provides a canonical metric in an orbit under $C_+^\infty(M)$ of conformally flat metrics. We owe the theorem to S.Y. Cheng. Its proof will appear elsewhere. Of course M is always a compact hyperbolic manifold in what follows.

Theorem 8.1. Every orbit under $C_+^\infty(M)$ of conformally flat metrics contains a metric of constant scalar curvature. The metric is unique up to scalar multiples.

We will use two different normalizations of the scalar.

Corollary 1 (first normalization). Every orbit under $C_+^\infty(M)$ of conformally flat metrics contains a unique metric g of constant scalar curvature $-n(n-1)$.

Corollary 2 (second normalization). Every orbit under $C_+^\infty(M)$ of conformally flat metrics contains a unique metric g' of constant scalar curvature such that the volume of M (using the volume element associated to g') is 1.

Remark. The first corollary is the generalization of the theorem stating that every complex (conformal) structure on M^2 contains a unique hyperbolic metric.

Corollary 1 allows us to define an interesting function

$$\mathrm{vol}: C(M) \to \mathbb{R}_+$$

as follows. Let $c \in C(M)$ and g be the canonical metric with the first normalization. Then $\mathrm{vol}(c)$ is by definition the volume of M for the metric g. We can now relate the two normalizations g, g' in a conformal structure c, namely:

$$g' = \left(\frac{1}{\mathrm{vol}(c)}\right)^{2/n} g.$$

We now define a function $A: C(M) \to \mathbb{R}$ closely related to vol but more convenient for computations by:

$$A(c) = \int_M \tau(g')\mathrm{vol}'$$

Here $\tau(g')$ is the scalar curvature of g'. Since $\tau(\lambda g) = 1/\lambda\tau(g)$ for λ a positive constant we find:

$$A(c) = -n(n-1)(\mathrm{vol}(c))^{2/n}.$$

Before studying the function vol further, we point out another consequence of Theorem 8.1, the existence of a Petersson-Weil metric on $C(M)$. Now a Petersson-Weil metric on a space of structures is a consequence of a canonical metric in each structure and a Hodge theorem representing the infinitesimal deformations by "harmonic" tensor fields on M (as opposed to cohomology classes of tensor fields). The required Hodge Theorem has been proved by Gasqui and Goldschmidt [11].

We now prove some properties of the function vol. Of course, in the case $n = 2$, the function vol is constant by the Gauss-Bonnet Theorem. For all n, the unique hyperbolic structure, to be denoted c_0, is a critical point of A, Berger [3], page 29, hence, a critical point of vol. That the situation for $n > 2$ is altogether different from that of $n = 2$ is clear from the following theorem.

Theorem 8.2. *If* $n \geq 3$ *the second derivative of* $\mathrm{vol}: C(M^n) \to \mathbb{R}_+$ *at the hyperbolic structure* c_0 *is positive definite. In particular* vol *is not constant on* $C(M)$ *provided* $n \geq 3$.

Proof. The statement of the theorem is equivalent to the statement that the second derivative of A at the hyperbolic structure is negative definite. But the theorem now follows from [12], Theorem 2.5.

Indeed, we have only to check the eigenvalues of the operator L of [12], associated to the curvature transformation of the hyperbolic metric, on traceless symmetric 2-tensors. These eigenvalues are easily seen to be 0 on $e_i \otimes e_i - e_j \otimes e_j$ and -1 on $1/2(e_i \otimes e_j + e_j \otimes e_i)$. The minimum eigenvalue -1 is greater than $\min\{\frac{\tau}{n}, -\frac{\tau}{2n}\} = -(n-1)$ provided $n \geq 3$. With this the theorem is proved.

In the case in which $n = 4$ we find a remarkable and suggestive result using the Gauss-Bonnet Theorem.

Theorem 8.3. If $n = 4$; the function $\mathrm{vol}:\mathcal{C}(M^4) \to R_+$ has an absolute minimum at the hyperbolic structure c_0.

Proof. Let c be a conformal structure on M i.e. a canonical metric with the first normalization. In Berger [2], there is a formula for the Gauss-Bonnet integrand B as a universal linear combination of the norm $\|R\|^2$ of the curvature transformation R, the norm $\|Ric\|^2$ of the Ricci transformation Ric and τ^2 the square of the scalar curvature. For a conformally flat manifold R is a linear function of Ric so B must be a universal combination of $\|Ric\|^2$ and τ^2. By computing for S^4 and $S^1 \times S^3$ we find:

$$B = -2\|Ric\|^2 + \frac{2}{3}\tau^2.$$

Hence

$$\frac{2}{3}\int_M \tau^2 = 32\pi^2 \chi(M) + 2\int_M \|Ric\|^2$$

By Cauchy-Schwarz, we have for a symmetric transformation S:

$$(\mathrm{tr}\ S)^2 \leq \|S\|^2 n$$

if $(\lambda_1, \lambda_2, \ldots, \lambda_n)$ are the eigenvalues we have:

$$((\lambda_1, \lambda_2, \ldots, \lambda_n), (1, 1, \ldots, 1))^2 \leq (\lambda_1^2 + \cdots + \lambda_n^2)n.$$

We obtain then:

$$\tau^2 \leq \|Ric\|^2 4$$

Substituting

$$\frac{2}{3}\int_M \tau^2 \geq 32\pi^2 \chi(M) + \frac{1}{2}\int_M \tau^2$$

and:

$$\int_M \tau^2 \geq 6(32\pi^2)\chi(M)$$

By the Hirzebruch Proportionality Principle we have:

$$\chi(M) = 2\frac{\mathrm{vol}(c_0)}{\frac{8\pi^2}{3}} = \frac{6\ \mathrm{vol}(c_0)}{8\pi^2}$$

where $vol(c_0)$ denotes the volume of M for the hyperbolic metric. Hence:

$$\int_M \tau^2 \geq 144 vol(c_0).$$

But the canonical metric is normalized so that $\tau = -12$. We obtain:

$$\int_M \tau^2 = 144 \text{ vol } M = 144 \text{ vol}(c)$$

and hence

$$vol(c) \geq vol(c_0).$$

With this the theorem is proved.

References

[1] B.N. Apanasov, Nontriviality of Teichmuller space for Kleinian group in space, Riemann Surfaces and Related Topics, Proceedings of the 1978 Stony Brook Conference, Annals of Math. Studies No. 97, Princeton University Press (1980), 21-31.

[2] M. Berger, P. Gauduchon and E. Mazet, Le Spectre d'une Variete Riemanniene, Lecture Notes in Mathematics, 194, Springer-Verlag, New York.

[3] M. Berger, Quelque formules de variation pour une structure Riemanniene, Ann. Scient. Ec. Norm. Sup., 4e serie, t·3 (1970), 285-294.

[4] D. Birkes, Orbits of linear algebraic groups, Annals of Math. 93 (1971), 459-475.

[5] A. Borel, Compact Clifford-Klein forms of symmetric spaces, Topology 2 (1963), 111-122.

[6] A. Borel and N. Wallach, Continuous Cohomology, Discrete Subgroups, and Representations of Reductive Groups, Annals of Math. Studies No. 94, Princeton University Press (1980).

[7] S.S. Chen and L. Greenberg, Hyperbolic Spaces, Contributions to Analysis, A Collection of Papers Dedicated to Lipman Bers, Academic Press (1974), 49-87.

[8] P. Cohen, Decision procedures for real and p-adic fields, Comm. Pure Appl. Math., 22 (1969), 131-135.

[9] S.P. Eilenberg and S. MacLane, Cohomology theory in abstract groups I, Annals of Math. 48 (1947), 51-78.

[10] J. Gasqui and H. Goldschmidt, theoremes de dualite en geometrie conforme I and II, preprints.

[11] N. Koiso, On the second derivative of the total scalar curvature, Osaka Journal 16 (1979), 413-421.

[12] B. Kostant, The principal three-dimensional subgroup and the Betti numbers of a complex simple Lie group, Amer. J. of Math 81 (1959), 973-1032.

[13] W.L. Lok, Deformations of locally homogeneous spaces and Kleinian groups, thesis, Columbia University (1984).

[14] J. Millson, On the first Betti number of a constant negatively curved manifold, Annals of Math. 104 (1976), 235-247.

[15] J. Millson and M.S. Raghunathan, Geometric construction of cohomology for arithmetic groups I, Geometry and Analysis, Papers Dedicated to the Memory of V.K. Patodi, Springer (1981), 103-123.

[16] J. Morgan, Group actions on trees and the compactification of the spaces of classes of SO(n,1)-representations, preprint.

[17] D. Mumford and J. Fogarty, Geometric Invariant Theory, Ergenbnisse der Mathematik und ihrer Grenzgebiete 34, Springer (1982).

[18] O.T. O'Meara, Introduction to Quadratic Forms, Die Grundlehren der Mathematischen Wissenschaften, 117, Springer (1963).

[19] P.E. Newstead, Introduction to Moduli Problems and Orbit Spaces, Tata Institute Lecture Notes, Springer (1978).

[20] R. Palais, On the existence of slices for actions of non-compact Lie groups, Annals of Math. (2) 73 (1961), 295-323.

[21] M.S. Raghunathan, Discrete Subgroups of Lie Groups, Ergebnisse der Mathematik und ihrer Grenzgebiete 68, Springer (1972).

[22] M.S. Raghunathan, On the first cohomology of discrete subgroups of semi-simple Lie groups, Amer. J. Math. 87 (1965), 103-139.

[23] J.P. Serre, Trees, Springer (1980).

[24] D. Sullivan, Discrete conformal groups and measurable dynamics, Bull of the American Math. Soc. (new series) 6 (1982), 57-73.

[25] W.P. Thurston, The Geometry and Topology of Three-Manifolds, Princeton University Lecture Notes.

[26] V.S. Varadarajan, Harmonic Analysis on Real Reductive Groups, Lecture Notes in Mathematics 576, Springer.

[27] C. Kourouniotis, Deformations of hyperbolic structures on manifolds of several dimensions, thesis, University of London, 1984.

[28] W. Goldman and J. Millson, Local rigidity of discrete groups acting on complex hyperbolic space. To appear in Inv. Math.

[29] R. Schoen, Conformal deformations of a Riemannian metric to constant scalar curvature, J. Differential Geometry 20 (1984), 479-495.

[30] R. Zimmer, Ergodic Theory and Semisimple Groups, Monographs in Mathematics, Birkhauser, 1984.

DEPARTMENT OF MATHEMATICS
UNIVERSITY OF CALIFORNIA
LOS ANGELES, CA 90024

ON DIVISION OF FUNCTIONS, SOLUTION OF MATRIX EQUATIONS, AND PROBLEMS IN DIFFERENTIAL GEOMETRY AND PHYSICS

by Mark Alan Mostow

Dedicated to my father on his sixtieth birthday

In this article we present some results on the continuity of division of smooth functions and discuss their applications to linear algebra, differential geometry, and physics. Much of the work was done jointly with Steven Shnider and will appear in greater detail elsewhere ([MS2], [MS3], [Mos2].

The basic division problem treated here, which we shall call the joint continuity of division of smooth functions, is the following:

> Consider the collection of triples (f,g,h) of smooth (C^∞) real-valued functions on $\mathbb{R}^n$, or more generally, on a manifold M, satisfying the relation $f = g \cdot h$ (product). Assuming that $g^{-1}(0)$ is nowhere dense (i.e. that its complement is dense), we can write $h = f/g$ without ambiguity. Is the quotient h a (jointly) continuous function of the pair $(f = gh, g)$, with respect to the Fréchet C^∞ topology of uniform convergence of a function and its derivatives on compact sets?

This question appears not to have been considered explicitly. What has been studied is the continuity in the numerator of division by a fixed smooth function g, that is, of the operator sending $f = gh$ to $h = f/g$. For example, Łojasiewicz [Loj] proved that division by a real analytic function is continuous. We refer to [MS2] for a discussion of the problem of continuity in the numerator and its relation to closedness of ideals in rings of smooth functions and to divisibility of distributions by smooth functions; see [Horm] for its relation to the existence of tempered solutions of partial differential equations. But continuity in the numerator does not imply joint continuity, as the following simple example demonstrates. Let $M = \mathbb{R}$ with coordinate x, and let c be a real parameter. For each value of c, let $f_c(x) = c$ and $g_c(x) = x^2 + c^2$. For every c, h_c = (def.) $f_c/g_c = c/(x^2 + c^2)$ is a well-defined smooth function. Now h_o is the zero function, but h_c does not approach zero uniformly as c approaches 0, since $h_c(0) = c/c^2 \to \infty$ as $c \to 0$. Thus division is not jointly continuous at the pair $(f_o, g_o) = (0, x^2)$, even though division by g_c is continuous in the numerator for each c.

Though the joint continuity problem is a basic question about division of functions, our interest in it arose because of its appearance in a chain of problems starting with a problem in physics. This chain of problems will be discussed in Section 1. In Section 2 we give a precise statement of our result on joint continuity (Theorem 1), show why it is inadequate for the desired applications, and state a more general theorem (Theorem 2) which suffices for the applications. The applications are then stated. Section 3 gives a proof of Theorem 1 and a discussion of the ideas of the proof of Theorem 2, whose details will appear in a joint paper with Steven Shnider [MS2].

1. Origin of the joint continuity of division problem

The evolution of our work leading to the joint continuity problem can be described schematically as follows:

Physics → Differential geometry →
Linear algebra → Division problem.

Physics.

A subject of intense activity by physicists and mathematicians in recent years has been gauge field theory, which seeks to explain the fundamental forces of nature at the subnuclear level. By analogy with the electromagnetic potentials and fields used to describe electromagnetic interactions, one describes subnuclear interactions by means of gauge potentials and gauge fields, which have physical meaning after one imposes the equivalence relation induced by an infinite-dimensional group of gauge transformations. Each gauge potential determines a corresponding gauge field, but the map from potentials to fields is not one-to-one, even at the level of equivalence classes. This lack of injectivity is called the field copy phenomenon by physicists and was studied by the author in [Mos1].

Each potential contains the information necessary to describe a physical state. In the functional integral approach to gauge field theory [Gli J], one computes the values of certain physical quantities by taking weighted averages over the space of gauge potentials, using functional integrals. (We ignore here the thorny problem of making mathematical sense of these integrals.) A related approach, advocated by Halperin [Hal] and others, is to integrate instead over the space of gauge fields. This has some physical and mathematical advantages. For example, a gauge field is analogous to an electromagnetic force field, which is more directly observable than a potential. Also, the group of gauge transformations acts linearly and tensorially on the space of gauge fields, but only affinely and non-tensorially on the space of gauge potentials.

In an attempt to understand the mathematical relation between these two uses of functional integrals in gauge field theory, Steven Shnider and the author studied the relation between spaces of gauge potentials and of gauge fields [MS1], endowed with function space topologies. Though, as mentioned, the map from potentials to fields is not one-to-one, there is an open dense set of potentials which does map one-to-one into the space of gauge fields, under certain hypotheses which are commonly satisfied. We showed, however, that the inverse of this restricted map is not continuous. We were left with the question:

> Is there an open dense, or at least generic (in the sense of [Gol G]) subset of the space of gauge potentials which maps one-to-one <u>and</u> <u>homeomorphically</u> onto a subset of the gauge fields?

A positive answer to this question might help show that integrating over gauge potentials is mathematically equivalent to integrating over gauge fields. Of course, one would still have to compare measures on the two spaces.

<u>Differential geometry</u>.

The preceding physical problem becomes a mathematical one when we use the gauge field theory dictionary:

gauge potential = connection of a principal bundle

gauge field = curvature of the connection

We make the restriction throughout that the base space of the principal bundle is 4-dimensional and that the structure group G of the bundle is semi-simple. The physical question becomes a question in differential geometry:

> Is there a generic set of connections which depend uniquely and continuously on their curvatures, in suitable function space topologies?

A connection μ is a 1-form on the total space of the bundle taking values in the Lie algebra $\mathfrak{g}$ of G, which is ad_G-equivariant and restricts to the Maurer-Cartan form on each fiber [Kob N]. Its curvature Ω is the 2-form

$$\Omega = d\mu + \frac{1}{2}[\mu \wedge \mu],$$

where $[\cdot\wedge\cdot]$ is the wedge product of $\mathfrak{g}$-valued forms with respect to the bilinear form $\mathfrak{g} \times \mathfrak{g} \to \mathfrak{g}$ defined by the bracket product of $\mathfrak{g}$.

Our problem can be approached by using the Jacobi identity

$$d\Omega = [\Omega \wedge \mu].$$

Let ad_Ω denote the linear map $\tau \to [\Omega \wedge \tau]$ from $\mathfrak{g}$-valued 1-forms to $\mathfrak{g}$-valued 3-forms which are ad_G-equivariant and restrict to 0 on each fiber. At each point, each of these forms corresponds to an element of $\mathrm{Hom}(\mathbb{R}^4, \mathfrak{g})$, a vector space of dimension $4 \dim G$, so we may represent ad_Ω at each point by a square $4 \dim G \times 4 \dim G$ matrix. If μ_o is an arbitrary "reference" connection we may write

$$d\Omega = [\Omega \wedge \mu_o] + ad_\Omega(\mu - \mu_o).$$

By abuse of notation we write

$$d\Omega = ad_\Omega(\mu).$$

Inverting this, we get a <u>formal</u> relation

$$\mu = ad_\Omega^{-1}(d\Omega)$$

for the connection as a function of the curvature.

In [MS1] we exhibited open dense subsets A_n of the space of connections, each consisting of connections for which ad_Ω^{-1} exists at all points in an open dense subset of the total space of the bundle. The operator F taking μ to Ω must be injective on each A_n, since each $\mu \in A_n$ is uniquely determined by Ω on a dense subset of the total space, and hence everywhere, by continuity. For the specific sets A_n we proved that $(F|A_n)^{-1}$ is not continuous. Nor did we find, in [MS1], another generic set A' for which $(F|A')^{-1}$ is continuous.

By taking another approach, however, we shall find such a set. We begin by ignoring the extraneous geometric structure and reducing to a simpler problem in linear algebra.

Linear algebra. Consider triples (A,X,B) satisfying $AX = B$. Here A,X,B are smooth (C^∞) functions on $\mathbb{R}^m$, with X and B taking their values in the vector space $\mathbb{R}^n$, and A in the real $n \times n$ matrices. (We have in mind $A = ad_\Omega$, $X = \mu$, $B = d\Omega$.) One can show that for a generic set of matrix-valued functions A, the inverse A^{-1} exists on an open dense subset of $\mathbb{R}^m$. On the latter subset, and hence, by continuity, on all of $\mathbb{R}^m$, X is uniquely determined by $X = A^{-1}B$. But does X depend continuously on (A,B) (in a function space topology), for some possibly smaller generic set of matrix-valued functions A?

We study this question by multiplying both sides of $AX = B$ by C_A, the transpose of the matrix of cofactors of A. This yields

$$(\det A)X = C_A B.$$

Hence formally we have

$$X = \frac{C_A B}{\det A} .$$

But now we see that the continuity of the operator $(A,B) \to X$ would follow from the joint continuity of division by the function $\det A$. Thus we have arrived at the division problem discussed above.

2. Results

In this section we present two results on joint continuity of division of smooth functions and indicate some applications.

Let M be a smooth (C^∞), n-dimensional (n finite), second countable, Hausdorff manifold. Let $E = E(M)$ be the Frechét space of smooth functions from M to $\mathbb{R}$, with Frechét C^∞ topology defined (as usual) by the semi-norms

$$\|f\|_{r,K,u} = \sup_{x\in K}\sup_{0\leq|A|\leq r}|D^A f(x)|,$$

where $K \subset M$ is compact, $0 \leq r < \infty$, $A = (k_1,k_2,\ldots,k_n)$ is a multi-index of order $|A| = k_1 +\cdots+ k_n$, and $D^A = d_1^{k_1},\ldots,d_n^{k_n}$ ($d_m = d/dx_m$) is the corresponding partial derivative operator, relative to some smooth coordinate chart $u = (x_1,\ldots,x_n)$ whose domain includes K. (We shall generally suppress the u in the notation and write simply $\|f\|_{r,K}$.)

Theorem 1 says that the division map taking $(f = gh,g)$ to $h = f/g$ is jointly continuous near any pair (gh,g) for which 0 is not a critical value of g. More precisely:

Theorem 1. Let

$$S = \{g \in E \mid 0 \text{ is not a critical value of } g\}$$
$$= \{g \mid \text{the differential } dg \neq 0 \text{ at all } x \text{ where } g(x) = 0\}.$$

Let $\mathfrak{m}$ be the multiplication map

$$\mathfrak{m}: E \times S \to E \times S$$

defined by

$$\mathfrak{m}(h,g) = (gh,g).$$

Then $\mathfrak{m}$ is a topological embedding. That is, $\mathfrak{m}$ is continuous and one to one, and the division map

$$\mathfrak{m}^{-1}: (gh,g) \mapsto (gh/g,g) = (h,g)$$

from image($\mathfrak{m}$) to $E \times S$ is continuous. In particular, $(gh,g) \mapsto h$ is continuous on image($\mathfrak{m}$).

Remarks. 1) The theorem would be false if S were replaced by $\{g \mid$ some second derivative of g is non-zero wherever g and dg vanish$\}$, since division is not jointly continuous at $(0,x^2)$, as we have seen.

2) The difficulty in proving joint continuity of division clearly depends on the behavior of the denominators g near their zero sets. The hypothesis of the theorem guarantees not only that the zero sets are "nice", in fact, submanifolds, but also that they change continuously as g is varied. Also, each g looks like a coordinate function near its zero set. In the example $c/(x^2 + c^2)$, on the other hand, the zero set of $x^2 + c^2$ changes abruptly from $\{0\}$ when $c = 0$ to the empty set when $c \neq 0$.

3) Transversality theory ([Gol G], Chap. II, Sect. 4) implies that S is open and dense in $C^\infty(M)$ in the Whitney C^∞ topology (which is finer than the Fréchet C^∞ topology on E), or even in the Whitney C^1 topology, since S can be described as the set of functions $f \in C^\infty(M)$ which are transversal to the zero-dimensional submanifold $\{0\} \subset \mathbb{R}$. In the Fréchet C^∞ topology, therefore, S is dense, but it is not open unless M is compact, in which case the two topologies coincide.

4) We have stated the theorem for the Fréchet rather than the Whitney C^∞ topology because the Fréchet C^∞ topology measures the closeness of functions locally, and the division problem is local, too. Even for the problem of continuity of division by a fixed function there are global conditions that must be satisfied for division to be continuous in the Whitney C^∞ topology (see [MS2]).

Unfortunately, Theorem 1 is not strong enough to answer our original questions. For example, in the geometric problem we have to divide by the function $\det(\mathrm{ad}_\Omega)$. When the Lie algebra is $\mathfrak{so}(3)$, a result of Wu and Yang shows that this determinant is a perfect square (see [MS1]) and hence has gradient zero on its whole zero set. While in this case we can get around the difficulty by dividing twice by the square root of $\det(\mathrm{ad}_\Omega)$, it seems hard to guarantee that for more complicated Lie algebras the function $\det(\mathrm{ad}_\Omega)$ can always be factored into functions which generically satisfy the hypothesis of Theorem 1. What is more, Remark 1 above shows that there is virtually no hope of weakening the hypothesis of Theorem 1. Instead, we state a more technical division theorem which will solve our problems.

Theorem 2. [MS2] Let $p:P \to \mathbb{R}$ be a fixed analytic function on a real-analytic manifold P. Let $\mathcal{E}(M)$ (respectively $\mathcal{E}(M,P)$) denote the Fréchet space of C^∞ functions from a smooth manifold M to $\mathbb{R}$ (resp. from M to P), with the Fréchet C^∞ topology. Let $\mathbb{M}_p$ be the operator

$$\mathbb{M}_p:\mathcal{E}(M) \times \mathcal{E}(M,P) \ni (h,G) \mapsto (f,G) \in \mathcal{E}(M) \times \mathcal{E}(M,P),$$

where $f(x) = h(x)p(G(x))$. Choose a Whitney stratification of $p^{-1}(0)$ for which the vanishing order of p in P is constant on each stratum. Let W be the set of all $G \in \mathcal{E}(M,P)$ which are transversal to every stratum. (W is open and dense in the Whitney C^1 topology.) Then $\mathbb{M}_p|\mathcal{E}(M) \times W$ is one-to-one, and its inverse operator, defined on $\mathbb{M}_p(\mathcal{E}(M) \times W)$ and mapping (f,G) to $((f/(p\circ G)),G)$, is continuous (in the Fréchet C^∞ topology). In particular, $(f,G) \to f/(p\circ G)$ is continuous.

Remarks.

1) By letting p be the identity map of $\mathbb{R}$, we obtain Theorem 1 as a corollary of Theorem 2.

2) It is crucial here that the quotient $f/p\circ G$ depends jointly continuously not on the numerator f and denominator $p\circ G$, but rather on the pair (f,G).

Theorem 2 applies immediately to the problem of whether X depends continuously on (A,B) in the equation $B = AX$ of vector-valued smooth functions. We obtained a formal relation $X = (C_A B)/(\det A)$. To apply Theorem 1 we would have to know that $\det A$ does not have zero as a critical value, but we can apply Theorem 2 without such a condition by choosing p to be the polynomial $\det:\mathbb{R}^{n^2} \to \mathbb{R}$. Doing so, we obtain:

Corollary. Let $\mathcal{X}$ be the space of smooth (C^∞) $\mathbb{R}^n$-valued functions on a smooth m-manifold M, $\mathcal{A}$ the space of smooth $(n \times n)$-matrix-valued functions on M, and

$$\mathcal{A}_0 = \{A \in \mathcal{A} \mid \det A \neq 0 \text{ on a dense set in } M\}.$$

Endow them with the Fréchet C^∞ topology. Then the map

$$\mathbb{M}: \mathcal{A}_0 \times \mathcal{X} \ni (A,X) \to (A,AX) \in \mathcal{A}_0 \times \mathcal{X}$$

is one-to-one, and the inverse map $\mathbb{M}^{-1}$: image $\mathbb{M} \to \mathcal{A}_0 \times \mathcal{X}$ is continuous at all pairs (A,AX) for which A is transversal to a Whitney stratification of $\det^{-1}(0) \subset \mathbb{R}^{n^2}$, provided that the vanishing order of det in $\mathbb{R}^{n^2}$ is constant on each stratum.

Proof. At each $x \in M$, $(\det A)X = C_A(AX)$ (Cramer's Rule), where C_A is the transpose of the matrix of cofactors of A. Now apply Theorem 2, taking h = ith entry of X, f = ith entry of $C_A(AX)$, $G = A$, and $p = \det$.

Q.E.D.

Remark. The set of all A satisfying the hypotheses is open and dense in the Whitney C^∞ topology and dense in the Fréchet topology. Hence generically, X depends uniquely and continuously on (A,AX).

At last we consider our original physical/geometric problem. The last result about linear algebra can be used to show that for a principal bundle over space-time with semi-simple structure group, there exists a generic set of connections (gauge fields) with the desired property. The proof contains some subtleties and will appear elsewhere [MS3].

As a final note we speculate that Theorem 2 may say something about the existence of continuous families of fundamental solutions of parametrized families of partial differential equations.

3. Proofs.

Although Theorem 1 is a corollary of Theorem 2, it is instructive to prove Theorem 1 directly using only elementary concepts.

Synopsis of proof of Theorem 1.

First we reduce to proving continuity near pairs $(f,g) = (gh,g)$ with $f = 0 = h$. By the Implicit Function Theorem we can use coordi-systems with g as the first coordinate. Write

$$f(g,x_2,\ldots,x_n) = \int_0^1 (d/dt)(f(tg,x_2,\ldots,x_n))dt$$
$$= g \int_0^1 (d_1 f)(tg,x_2,\ldots,x_n)dt.$$

From this, we get uniform bounds for $h = f/g$ locally in terms of bounds on $d_1 f$, and for the r-jet $j^r h$ of h in terms of bounds on $j^{r+1}f$, for each r, relative to the $(g,x_2,\ldots,x_n)$ coordinate system. When we change to a coordinate system independent of g, we get uniform bounds on compact sets for $j^r h$ in terms of similar bounds on $j^{r+1}f$ and $j^{r+1}g$, locally in $\mathcal{E} \times \mathcal{S}$. This implies the result.

Detailed proof of Theorem 1.

The continuity of $\mathfrak{m}$ is an immediate consequence of Leibniz' formula for derivatives of a product. The injectivity of $\mathfrak{m}$ follows from the fact that g is in $\mathcal{S}$, implying that the complement of $g^{-1}(0)$ is dense and hence that h is uniquely determined by gh and g. To show that m is an embedding, it thus suffices to prove that if $g_k \to g_0$ in $\mathcal{S}$, and $\{h_k\}$ is a sequence in $\mathcal{E}$ such that $g_k h_k$ converges to $g_0 h_0$ for some h_0 in $\mathcal{E}$, then $h_k \to h_0$ in $\mathcal{E}$. We can reduce to the case $h_0 = 0$, because if that case were proven, we would have the implications

$$g_k h_k \to g_0 h_0 \Rightarrow$$
$$g_k(h_k - h_0) + h_0(g_k - g_0) \to 0 \Rightarrow$$
$$g_k(h_k - h_0) \to 0 \Rightarrow h_k - h_0 \to 0.$$

It will therefore suffice to prove the following two lemmas. We also use the trivial fact that if K is a given compact subset of $\mathbb{R}^n$ contained in the union of compact subsets $K_1,\ldots,K_m$, then

$$\|\,\|_{r,K} \le \max_{j=1,\ldots,m} \|\,\|_{r,K_j}.$$

Lemma 1. Let $M = \mathbb{R}^n$. For every function $g_0 \in S$ and every point $x_0 \in \mathbb{R}^n$ for which $g_0(x_0) \neq 0$, there exist positive constants S and B_r, $r = 0,1,2,\ldots$, and a compact neighborhood K of x_0 such that

$$\|h\|_{r,K} \le B_r \|f\|_{r,K} \max[1, (\|g\|_{r,K})^r]$$

for all $f, h \in E$ and $g \in S$ satisfying $f = gh$ and $\|g-g_0\|_{r,K} < \delta$.

Lemma 2. Let $M = \mathbb{R}^n$. For every function $g_0 \in S$ and every point $x_0 \in R^n$ for which $g_0(x_0) = 0$, there exist positive constants δ and B'_r, $r = 0,1,2,\ldots$, and a compact neighborhood K of x_0 such that

$$\|h\|_{r,K} \leq B'_r \|f\|_{r+1,K} \max[1, (\|g\|_{r+1,K})^{(r^3-r^2+4r+1)}]$$

for all $f, h \in E$ and $g \in S$ satisfying $f = gh$ and $\|g-g_0\|_{1,} < \delta$.

Proof of Lemma 1. If $g_0(0) \neq 0$, then we can choose a compact neighborhood K of x_0 and find positive constants δ and ε so that $|g| \geq \varepsilon$ on K whenever $\|g-g_0\|_{0,K} < \delta$. We assume, for the sake of convenience, that $\varepsilon \leq 1$. For such functions g we can write $h = f/g = f\cdot(1/g)$ on K. By elementary calculus, therefore, any partial derivative $D^A h$ of h of order $|A| \leq r$ is a sum of terms, each of which is a product of a coefficient, a derivative of f of order $\leq r$, up to r derivatives of g, each of order $\leq r$, and a power of $(1/g)$ between the first and the $(r+1)$st, inclusive. It follows that the absolute value of each term is bounded by

$$C\|f\|_{r,k}\{\max[1, (\|g\|_{r,K})^r]\} / \varepsilon^{r+1},$$

where C is a constant which depends on A and on the term but not on f, g, or h. By adding these inequalities and taking the maximum over all multi-indices A with $|A| \leq r$, we obtain an inequality

$$\|h\|_{r,K} \leq B_r \|f\|_{r,K} \max[1, (\|g\|_{r,K})^r],$$

when $\|g-g_0\|_{0,K} \leq \delta$, for some positive constant B_r which depends on ε.

Proof of Lemma 2. The hypotheses imply that $dg_0 \neq 0$ at x_0. Without loss of generality we may assume that $d_1 g_0 > 0$ at x_0. Let K be a compact rectangular neighborhood of x_0 in $\mathbb{R}^n$, defined by

$$K = \{x = (x_1, x_2, \ldots, x_n) \in \mathbb{R}^n | a_m \leq x_m \leq b_m,\ m = 1,2,\ldots,n\}.$$

Fix a positive constant $\varepsilon < \min[1, (d_1 g_0)(x_0)]$. Assume that K is chosen small enough so that

$$d_1 g_0 > \varepsilon \quad \text{on} \quad K. \tag{*}$$

The implicit Function Theorem [Gol G, pp. 7-8] implies that near x_0, $g_0^{-1}(0)$ is the graph of some smooth function $x_1 = G_0(x_2,\ldots,x_n)$. By shrinking the intervals $[a_2,b_2],\ldots,[a_n,b_n]$ if necessary, we can arrange, using the monotonicity of g_0 on each interval of constant $(x_2,\ldots,x_n)$ in K, that

$K \cap g_0^{-1}(0)$ is the graph of the restriction of G_0 to $\{(x_2,\ldots,x_n) \mid a_m \le x_m \le b_m$ for $m = 2,\ldots,n\}$, and x_1 never equals a_1 or b_1 on $K \cap g_0^{-1}(0)$. (**)

Choose δ small enough to guarantee that every function $g \in S$ with $\|g-g_0\|_1 = \text{(def.)}\ \|g-g_0\|_{1,K} < \delta$ also satisfies properties (*) and (**) (for the same K but different G instead of G_0. We shall henceforth suppress the K in $\|\ \|_{r,}$.). For example, one can satisfy (**) by choosing for δ the infimum of $|g_0(x)|$ on $K \cap \{x_1 = a_1$ or $b_1\}$. For all such functions g, the n-tuple of functions $x' = \text{(def.)}\ (x_1',x_2',\ldots,x_n')$, where $x_1' = g$ and $x_m' = x_m$, $m = 2,\ldots,n$, defines a C^∞ coordinate system on K (by the Inverse Function Theorem and direct verification that the map $x':K \to \mathbb{R}^n$ is one-to-one).

Let $K' = x'(K) \subset \mathbb{R}^n$. Define functions $f^\wedge$, $g^\wedge$, $h^\wedge$ from K' to $\mathbb{R}$ by $f^\wedge(x') = f(x)$, etc., where $x' = x'(x)$. Since $f = 0$ when $g = 0$,

$$\begin{aligned} f^\wedge(x_1',x_2',\ldots,x_n') &= \int_0^1 (d/dt)[f^\wedge(tx_1',x_2',\ldots,x_n')]dt \\ &= x_1' \int_0^1 (d_1'f^\wedge)(tx_1',x_2',\ldots,x_n')dt \end{aligned}$$

at all points $x' = (x_1',x_2',\ldots,x_n') \in K'$. (Here $d_1' = dx_1' = d/dg$). Hence

$$h(x) = h^\wedge(x') = f^\wedge/g^\wedge = \int_0^1 (d_1'f^\wedge)(tx_1',x_2',\ldots,x_n')dt$$

on K'.

We seek uniform bounds on K for h and its partial derivatives in terms of similar bounds for f and g. This is easy to do in the x' coordinate system, since we can differentiate under the integral sign to obtain

$$\begin{aligned}(D')^A h\hat{} &= (\text{def.})(d_1')^{k^1}(d_2')^{k^2}\cdots(d_n')^{k^n} h\hat{} \\ &= \int_0^1 t^{k^1}[d_1'(D')^A f\hat{}](tx_1', x_2', \ldots, x_n')\,dt,\end{aligned}$$

which implies the inequality

$$\sup_{x' \in K'} |(D')^A h\hat{}(x')| \leq \sup_{x' \in K'} |d_1'(D')^A f\hat{}(x')|$$

(since for $0 \leq t \leq 1$, $(tx_1', x_2', \ldots, x_n') \in K'$ when $x' \in K'$). Let $\|h\|_r'$ denote the semi-norm $\|h\hat{}\|_{r,K'}$ on $h\hat{}(x')$, relative to the x' coordinate system. Then we have

$$\|h\|_r' \leq \|f\|_{r+1}'. \tag{1}$$

For our purposes, however, we need bounds on derivatives in a coordinate system independent of g. To this end we shall now relate derivatives in the x and x' coordinate systems.

Trivially, we have $\|F\|_0' = \|F\|_0$ for any function F. The Chain Rule implies that

$$\begin{aligned}d_1 &= (d_1 g)d_1', \\ d_m &= d_m' + (d_m g)d_1' \quad \text{for } m = 2, \ldots, n.\end{aligned}$$

If we express a partial derivative

$$D^A h = d_1^{k^1} d_2^{k^2} \cdots d_n^{k^n} h \quad (1 \leq |A| \leq r)$$

in terms of the operators $d_1', d_2', \ldots, d_n'$, using these relations, we get

$$D^A h(x) = ((d_1 g)d_1')^{k^1}(d_2' + (d_2 g)d_1')^{k^2}\cdots(d_n' + (d_n g)d_1')^{k^n} h\ (x').$$

When we expand this using the product rule, we get a sum of terms, each a product of a coefficient, a partial derivative of $h\hat{}$ of order $\leq r$ with respect to $(x_1', x_2', \ldots, x_n')$, and a function which is a product of $\leq r$ derivatives of order $\leq r-1$ of $d_1g,\ldots,d_ng$ with respect to $(x_1',x_2',\ldots,x_n')$. In terms of norms, we get

$$\|h\|_r \leq C\|h\|_r'\max[1,(\|dg\|_{r-1}')^r] \tag{2}$$

for some positive constant C, where

$$\|dg\|_{r-1}' = \max_{m=1,\ldots,n}\|d_mg\|_{r-1}'.$$

Finally, we need a bound for x' derivatives in terms of x derivatives. Using the relations

$$d_1' = (1/d_1g)d_1,$$
$$d_m' = d_m - (d_mg)d_1' = d_m - (d_mg/d_1g)d_1 \quad \text{for} \quad m = 2,\ldots,n,$$

we obtain, by a computation similar to the preceding one,

$$\|F\|_r' \leq$$
$$C'\|F\|_r \max[1,\|1/d_1g\|_{r-1},\|d_2g/d_1g\|_{r-1},\ldots,\|d_ng/d_1g\|_{r-1}]^r \tag{3}$$

for any $F \in E$. By the proof of Lemma 1,

$$\begin{aligned}\|1/d_1g\|_{r-1} &\leq C_1\max[1,(\|d_1g\|_{r-1})^{r-1}]\\ &\leq C_1\max[1,(\|g\|_r)^{r-1}],\end{aligned} \tag{4}$$

and

$$\begin{aligned}\|d_mg/d_1g\|_{r-1} &\leq C_2\|d_mg\|_{r-1}\max[1,(\|d_1g\|_{r-1})^{r-1}]\\ &\leq C_2\|d_mg\|_{r-1}\max[1,(\|g\|_r)^{r-1}]\\ &\leq C_2\max[1,(\|g\|_r)^r]\end{aligned} \tag{5}$$

for positive constants C_1, C_2 depending on ε. Combining (3), (4), and (5), we get

$$\|F\|_r' \leq C''\|F\|_r\max[1,(\|g\|_r)^{r^2}]. \tag{6}$$

By (1), (2), and (6), in the case $r \geq 1$, we have

$$\begin{aligned}\|h\|_r &\le C\|h\|_r'\max[1,(\|dg\|_{r-1}')^r] \\ &\le C\|f\|_{r+1}'\max[1,(\|dg\|_{r-1}')^r] \\ &\le C_3\|f\|_{r+1}\max[1,(\|g\|_{r+1})^{(r+1)^2}]\cdot \\ &\qquad \max\{1,[\|dg\|_{r-1}\max(1,(\|g\|_{r-1})^{(r-1)^2})]^r\} \\ &\le C_3\ f_{\ r+1}\max[1,(\ g_{\ r+1})^{r^3-r^2+4r+1}]\end{aligned}$$

for f,g,h as above. The constants here depend on r, K, g_0, and ε. In the case $r = 0$ we get

$$\begin{aligned}\|h\|_0 = \|h\|_0' &\le \|f\|_1' \\ &\le C\|f\|_1\max[1,\|g\|_1],\end{aligned}$$

which is consistent with the general formula.

Q.E.D.

<u>Remarks</u>. We can use the proof of Theorem 1 to explain the essential difference between joint continuity of division and continuity (in the numerator) of division by a fixed denominator. Continuity of division by a fixed function g could be established by finding, for each $r = 0,1,2,\ldots$ and each compact set K, an integer r' (of necessity $\ge r$), a compact set K', and a positive constant C satisfying

$$\|h\|_{r,K} \le C\|gh\|_{r',K'} \quad \text{for all } h.$$

To establish <u>joint</u> continuity of division near $(f,g) = (0,g)$, we must find constants r', C that work for all denominators in some neighborhood of g in $\underline{\underline{E}}$, or equivalently, constants that depend continuously on g. Lemmas 1 and 2 do precisely that. For example, in Lemma 2 we get $r' = r+1$, $K' = K$ (for a specially constructed set K), and

$$C = B_r'\max[1,(\|g\|_{r+1,K})]^{r^3-r^2+4r+1},$$

which varies continuously as g varies in $\underline{\underline{E}}$.

The proof of Theorem 2 is based on similar ideas but is much harder because it involves stratified sets. A detailed proof will appear elsewhere ([MS2] and [Mos2]). Here we will just sketch the proof.

Sketch of Proof of Theorem 2. It suffices to find inequalities of the form

$$\|h\|_{r,K} \leq C\|f\|_{r',K'}$$

(K,K' compact, $K \subset K'$, $f = h\cdot(p\circ G)$, $\|h\|_{r,K}$ = supremum of $\|j^r h(x)\|$ over $x \in K$, using some norm on the fibers of $J^r(M)$)

in which C is independent of G locally in E(M,P). Hormander [Horm], in his proof that division by a polynomial is continuous, obtained bounds of this type for division of f by a fixed polynomial g (i.e. $f = gh$, $h = f/g$). The transversality hypothesis on G guarantees that the zero set of $g = (\text{def.})\ p\circ G$ varies "continuously" as G varies in E(M,P). Also, one can show [Mos2] that as x varies over a compact set in M, the distances $\text{dist}(x,g^{-1}(0))$ and $\text{dist}(G(x),p^{-1}(0))$ (using Riemannian metrics) are bounded by constant multiples of each other, and that these constants can be chosen to work for all maps G whose 1-jets are close enough to the 1-jet of a given map G_0 on a neighborhood of the compact set. Using these and other ideas, one can adapt Hörmander's proof carefully and prove that his constants can also be chosen to work for all G close enough to G_0 in E(M,P).

References

[Gli J] Glimm, J., Jaffe, A., Quantum Physics: A Functional Integral Point of View. New York: Springer, 1981.

[Gol G] Golubitsky, M., Guillemin, V., Stable Mappings and Their Singularities. New York: Springer-Verlag, 1973. Second, corrected, printing, 1980.

[Hal] Halpern, M.B., Field strength and dual variable formulations of gauge theory. Phys. Rev. D 19 (1979), pp. 517-530.

[Horm] Hörmander, L., On the division of distributions by polynomials. Arkiv for Matematik 3 (1958), pp. 555-568.

[Kob N] Kobayashi, S., Nomizu, K., Foundations of Differential Geometry, Part 1. New York: Interscience, 1963.

[Loj] Łojasiewicz, S., Sur le problème de la division. Studia Math. 18 (1959), pp. 87-136.

[Mos 1] Mostow, M.A., The field copy problem: to what extent do curvature (gauge field) and its covariant derivatives determine connection (gauge potential)? Commun. in Math. Phys. 78 (1980), pp. 137-150.

[Mos2] Mostow, M.A., Joint continuity of division of smooth functions II: The distance to a Whitney stratified set from a transversal submanifold. Trans. of the A.M.S. 292 (1985), pp.585-594.

[MS 1] Mostow, M.A., Shnider, S., Does a generic connection depend continuously on its curvature? Commun. in Math. Phys. 90 (1983), pp. 417-432.

[MS 2] Mostow, M.A., Shnider, S., Joint continuity of division of smooth functions. I. Uniform Łojasiewicz estimates. Trans. of the A.M.S. 292 (1985), pp. 573-583.

[MS 3] Mostow, M.A., Shnider, S., An application of a division theorem to the continuous determination of connections from curvatures. Preprint.

DEPARTMENT OF MATHEMATICS
NORTH CAROLINA STATE UNIVERSITY
RALEIGH, N.C. 27650

Current address:

CLARITY, LTD.
60 MEDINAT HAYEHUDIM STREET
P.O. BOX 3112
HERZELIA 46103
ISRAEL

STRONG RIGIDITY FOR KÄHLER MANIFOLDS AND THE CONSTRUCTION OF BOUNDED HOLOMORPHIC FUNCTIONS

by Yum-Tong Siu[1]

In this paper we shall give a survey of the known results and methods concerning the strong rigidity of Kähler manifolds and present some new related results. The important phenomenon of strong rigidity was discovered by Professor G.D. Mostow in the case of locally symmetric nonpositively curved Riemannian manifolds. He proved [18] that two compact locally symmetric nonpositively curved Riemannian manifolds are isometric up to normalization constants if they have the same fundamental group and neither one contains a closed one or two dimensional totally geodesic submanifold that is locally a direct factor. This last assumption is clearly necessary because of the existence of nontrivial holomorphic deformations of any compact Riemann surface of genus at least two. Mostow's result says that if one can rule out the possibility of contribution to the change of metric structure from certain submanifolds of dimension two or lower, the metric structure is rigidly determined by the topology for compact locally symmetric nonpositively curved manifolds. Mostow's result also holds for the noncompact complete case under the assumption of finite volume.

A question naturally arises whether the phenomenon of rigid determination by topology occurs for structures other than the metric structure for a suitable class of manifolds. S.T. Yau conjectured that one can replace the metric structure by the holomorphic structure and get strong rigidity for the class of compact Kähler manifolds of complex dimension at least two with negative sectional curvature. Yau's conjecture in its full generality is still open. In [26] a result on strong rigidity was proved which implies Yau's conjecture for the class of compact Kähler manifolds of complex dimension at least two whose curvature tensor satisfies a negativity condition called strong negativity, which is stronger than the condition of negative sectional curvature. The result of [26] is the following.

[1] Research partially supported by a National Science Foundation grant.

If a compact Kähler manifold of complex dimension at least two has strongly negative curvature, then any compact Kähler manifold which is homotopic to it must be either biholomorphic or antibiholomorphic to it.

Though as a pointwise condition the condition of strongly negative curvature is clearly stronger than the condition of negative sectional curvature, there is no known example of a compact complex manifold of complex dimension at least two which admits a Kähler metric with negative sectional curvature but does not admit a Kähler metric with negative sectional curvature but does not admit a Kähler metric whose curvature tensor is strongly negative. It is not easy to construct negatively curved compact Kähler manifolds of complex dimension at least two. Besides the locally symmetric ones, there are not many known examples of compact Kähler manifolds with negative sectional curvature even in complex dimension two. The first negatively curved compact Kähler surface whose universal cover is not biholomorphic to the ball was constructed by Mostow-Siu [21]. It was constructed by using an almost discrete automorphism subgroup of the two-ball generated by complex reflections, as investigated earlier by Mostow [19,20]. The same method can also be used to get such surfaces from the Deligne-Mostow [3] version of Picard's method [23] of constructing almost discrete automorphism subgroups of the two-ball. The surface constructed by Mostow-Siu [21] has a Kähler metric of strongly negative curvature.

The strong rigidity result for Kahler manifolds proved in [26] uses harmonic maps. If two Kähler manifolds are homotopy equivalent and if one of them has a Kähler metric of strongly negative curvature, the result of Eells-Sampson [7] implies that there is a map from the manifold without curvature condition to the strongly negatively curved manifold which is harmonic and which is also a homotopy equivalence. Then one uses a Bochner-type formula to conclude that the harmonic map must be either holomorphic or antiholomorphic.

The method of producing holomorphic objects by constructing harmonic ones first is a very powerful tool in several complex variables. The construction of a holomorphic object in the case of several complex variables requires in general the solution of an overdetermined system of equations, whereas to construct a harmonic object one only needs to solve an elliptic equation which is not overdetermined. Of course, in general, there is no way to conclude that the constructed harmonic object must be holomorphic unless there is some uniqueness result forcing harmonic objects to be holomorphic. For example, one obtains such uniqueness results by imposing some growth condition on the solution in the case of functions or some negative curvature condition on the target space in the case of maps between complex manifolds. Negative curvature conditions are in many ways related to the notion of boundedness, which should be thought of as a special type of growth condition. For instance, it is conjectured that on the universal cover of a compact negatively curved Kähler manifold there are enough bounded holomorphic functions to separate points and give local coordinates.

Earlier Lelong [13] used this method of finding holomorphic objects from harmonic ones to construct on the complex Euclidean space holomorphic functions with specified growth conditions whose divisor is a given complex-analytic hypersurface assumed to satisfy a suitable growth condition. He solved first an equation involving the Laplace operator and then used uniqueness results derived from the growth conditions to force the solution to satisfy an overdetermined system of equation.

The use of a Bochner-type formula to force a harmonic map to be holomorphic under suitable curvature conditions involves formulating the problem in terms of differential forms. One tries to prove that the harmonic form which is the (0,1)-differential of the map must be equal to the identically zero form; from this it follows that the map must be holomorphic. This result can be thought of as a quasilinear form of the vanishing theorem of Kodaira. In this method the curvature condition can be weakened to cover the case of compact quotients of irreducible bounded symmetric domains of complex dimension at least two [26,27,28]. Recently Jost-Yau [10,11] and Mok [17] refined the method and introduced holomorphic foliations to handle the case of irreducible compact quotients of polydiscs of complex dimension at least two. In this paper we are going to give an intrinsic interpretation of these holomorphic foliations that are associated to a harmonic map to a quotient of a polydisc. The pullback under the harmonic map of the component line bundles of the tangent bundle of the target manifold can be given a natural holomorphic structure. The (1,0)-differential of the map gives holomorphic 1-forms with coefficients in these line bundles over the domain manifold. The holomorphic foliations are defined by the kernels of these line-bundle-valued holomorphic 1-forms. By using this intrinsic interpretation of the holomorphic foliations and some other simple arguments, we present a more streamlined proof of the strong rigidity result of Jost-Yau and Mok. Also, we shall prove a result on the noncompact case with finite volume jointly obtained with Yau.

We also give in this paper a way of constructing bounded holomorphic functions on the universal covers of certain compact Kähler manifolds. Until now there has been no general method of constructing bounded holomorphic functions on a complex manifold. The main known methods for constructing holomorphic functions are the method of coherent analytic sheaves and the method of L^2 estimates in partial differential equations. Neither method can give a uniform bound on the holomorphic functions produced. To produce the desired bounded holomorphic functions on the universal cover, our method requires that the compact Kähler manifold admit a continuous map into some compact hyperbolic Riemann surface which is nonzero on the second homology group. We first use the theorem of Eells-Sampson to get a harmonic map homotopic to the given continuous map and then use the holomorphic foliation associated to the harmonic map to get a holomorphic map into another compact hyperbolic Riemann surface. By going to the universal covers of both the given Kähler manifold and this second Riemann surface, we get from the holomorphic map a nonconstant bounded holomorphic function on the universal cover. Unfortunately, this method is less useful than it appears at first sight. Though the condition that there should exist a continuous map that is nonzero on the second homology seems rather mild, it is very difficult to determine which compact Kähler manifolds satisfy such a condition. There seems to be no relationship between the existence of such a continuous map and any negative curvature condition.

Table of Contents

§1. Bochner Type Formula and Strong Rigidity.

(1.1) Let M and N be Riemannian manifolds and $f:M \to N$ be a smooth map. By the global energy $E(f)$ of f we mean the global L^2 norm over M of the differential df of f. The map f is said to be harmonic if it is a critical point for the global energy functional $E(f)$. The Euler-Lagrange equation for the functional $E(f)$ is simply that the Laplacian of f with respect to the Levi-Civita connections of M and N is zero. The first fundamental result for the existence of harmonic maps is the theorem of Eells-Sampson [7] which says that if M and N are compact and if the Riemannian sectional curvature of N is nonpositive, then there exists a smooth harmonic map in every homotopy class of maps from M to N. Hartman [9] later showed that if the Riemannian sectional curvature of N is negative, then there is only one harmonic map of rank ≥ 2 in each homotopy class of maps from M to N. When M and N are Kähler manifolds, every holomorphic map from M to N is harmonic.

(1.2) Assume that M and N are both compact Kähler manifolds of complex dimensions m and n respectively with m at least 2. Let $f:M \to N$ be a harmonic map. Let $h_{\alpha\bar{\beta}}$ be the Kähler metric of N, $\Gamma^{\alpha}_{\beta\gamma}$ be the Christoffel symbol and R with components

$$R_{\alpha\bar{\beta}\gamma\bar{\delta}} = \partial_{\alpha}\overline{\partial_{\beta}}h_{\gamma\bar{\delta}} - h^{\lambda\bar{\mu}}\partial_{\alpha}h_{\gamma\bar{\mu}}\overline{\partial_{\beta}}h_{\bar{\delta}\lambda}$$

be the curvature tensor of N. (Summation convention is used here and also in the rest of this paper except in certain cases when it is clear from the formula that the contrary is meant.) Straightforward direct computation yields a formula of Bochner type

$$\partial\bar{\partial}(h_{\alpha\bar{\beta}}\bar{\partial}f^{\alpha}\wedge\overline{\partial f^{\beta}}) = h_{\alpha\bar{\beta}}D\bar{\partial}f^{\alpha}\wedge\overline{D\bar{\partial}f^{\beta}} + R_{\alpha\bar{\beta}\gamma\bar{\delta}}\bar{\partial}f^{\alpha}\wedge\overline{\partial f^{\beta}}\wedge\partial f^{\gamma}\wedge\overline{\partial f^{\delta}}.$$

Here $\{f^{\alpha}\}$ is the representation of f in local coordinates and

$$D\bar{\partial}f^{\alpha} = \partial\bar{\partial}f^{\alpha} + \Gamma^{\alpha}_{\beta\gamma}\partial f^{\beta}\wedge\bar{\partial}f^{\gamma}$$

is the covariant derivative of $\bar{\partial}f$ in the $(1,0)$ direction with respect to the connection $\Gamma^{\alpha}_{\beta\gamma}$ of $h_{\alpha\bar{\beta}}$.

Let ω be the Kähler form of M. Then Stokes' Theorem applied to the above Bochner type formula yields

$$\int_M h_{\alpha\bar{\beta}}D\bar{\partial}f^{\alpha}\wedge\overline{D\bar{\partial}f^{\beta}}\wedge\omega^{m-2} + \int_M R_{\alpha\bar{\beta}\gamma\bar{\delta}}\bar{\partial}f^{\alpha}\wedge\overline{\partial f^{\beta}}\wedge\partial f^{\gamma}\wedge\overline{\partial f^{\delta}}\wedge\omega^{m-2} = 0.$$

The harmonicity of f means that the trace of the f^*T_N-valued (1,1)-form $D\bar{\partial}f$ with respect to the Kähler metric of M vanishes, where T_N is the holomorphic tangent bundle of N. It follows from simple multilinear algebra that

$$h_{\alpha\bar{\beta}} D\bar{\partial}f^{\alpha} \wedge \overline{D\bar{\partial}f^{\beta}} \wedge \omega^{m-2} = |D\bar{\partial}f|^2 \left(\frac{4}{m(m-1)}\omega^m\right)$$

where $|D\bar{\partial}f|^2$ means the pointwise L^2 norm of $D\bar{\partial}f$. Thus if

$$R_{\alpha\bar{\beta}\gamma\bar{\delta}} \bar{\partial}f^{\alpha} \wedge \overline{\partial f^{\beta}} \wedge \partial f^{\gamma} \wedge \overline{\partial f^{\delta}} \wedge \omega^{m-2}$$

equals a nonnegative multiple of the volume of M at every point of M, then both $D\bar{\partial}f$ and

$$R_{\alpha\bar{\beta}\gamma\bar{\delta}} \bar{\partial}f^{\alpha} \wedge \overline{\partial f^{\beta}} \wedge \partial f^{\gamma} \wedge \overline{\partial f^{\delta}} \wedge \omega^{m-2}$$

vanish identically on M. This leads us to the investigation of suitable negative curvature conditions under which the term involving R is necessarily nonnegative.

(1.3) The curvature tensor R defines naturally a Hermitian form H_R on $T_M \otimes \overline{T_M}$ by $H_R(X \otimes \bar{Y}) = R(X,\bar{Y},Y,\bar{X})$ for X, Y in T_M. For X, Y in T_M the Riemannian sectional curvature of M in the direction of $\operatorname{Re} X$ and $\operatorname{Re} Y$ is negative if and only if H_R is positive on $X \otimes \bar{Y} - Y \otimes \bar{X}$. Note that the opposite sign for the curvature and H_R is due to the chosen sign convention. We say that the curvature of M is strongly negative (respectively strongly seminegative) if H_R is positive (respectively semipositive) on all elements of $T_M \otimes \overline{T_M}$ of the form $X \otimes \bar{Y} + Z \otimes \bar{W}$, where X, Y, Z, W are elements of T_M. Clearly, strong negativity in this sense implies negativity of the sectional curvature.

From (1.2) and simple multilinear algebra we can conclude that if the curvature of M is strongly negative and the rank of f over $\mathbb{R}$ is at least 3 at some point, then f is either holomorphic or antiholomorphic. A holomorphic map between two compact Kähler manifolds which is a homotopy equivalence must be biholomorphic. Thus the following result is a consequence of the theorem of Eells-Sampson on the existence of harmonic maps. If a compact Kähler manifold of complex dimension at least two has strongly negative curvature, then any compact Kähler manifold which is homotopy equivalent to it must be biholomorphic or antibiholomorphic to it [26].

(1.4) One says that the bundle of (p,0)-forms of M is positive in the sense of Nakano [22] if the Hermitian form on $T_M \otimes \overline{\Lambda^p T_M}$ defined by R (as the generalization of H_R for the general case when p is not necessarily 1) is positive, where Λ^p means taking the p-fold exterior product. Moreover, we define the degree of nondegeneracy of the curvature of M, denoted by $d(M)$, as the maximum of $\dim_{\mathbb{C}} V + \dim_{\mathbb{C}} W$, where V and W are two orthogonal nonzero complex linear subspaces of T_M with the property that H_R vanishes on all elements of $T_M \otimes \overline{T_M}$ of the form $v \otimes \bar{w}$ with v from V and w from W.

The preceding argument with a more careful handling of the multilinear algebra actually yields the following: If the curvature of M is strongly seminegative and if the bundle of $(p,0)$-forms of M is positive in the sense of Nakano for some $p \geq d(M)$, then any harmonic map $f:M \to N$ which has real rank at least $2p+1$ at some point of M must be either holomorphic or antiholomorphic [28]. This, together with the theorem of Eells-Sampson, yields the following strong rigidity result for compact quotients of irreducible bounded symmetric domains of complex dimension at least two. Any compact Kähler manifold which is homotopy equivalent to a compact quotient of an irreducible bounded symmetric domain of complex dimension at least two must be biholomorphic or antibiholomorphic to it [26,27,28]. To see this, we use the following table which gives the degree d of nondegeneracy of the curvature of the various types of bounded symmetric domains. In all cases, the degree d is less than the complex dimension of the bounded symmetric domain, and the bundle of $(d,0)$-forms is positive in the sense of Nakano.

Type	Complex Dimension	Degree of Nondegeneracy
$I_{m,n}$	mn	$(m-1)(n-1)+1$
II_n	$n(n-1)/2$	$(n-2)(n-3)/2+1$
III_n	$n(n+1)/2$	$n(n-1)/2+1$
IV_n	n	2
V	16	6
VI	27	11

The degrees of nondegeneracy for the two exceptional domains were computed by Zhong [32].

§2. Holomorphic Structure of the Pullback of the Polydisc Tangent Bundle.

(2.1) Let M be a compact Kähler manifold, N be a compact quotient of the n-dimensional polydisc Δ^n by a discrete group G of biholomorphisms of Δ^n, and $f:M \to N$ be a harmonic map. We assume that every element of G is of the form $(z_1,\ldots,z_n) \to (g_1(z_1),\ldots,g_n(z_n))$, where each g_ν $(1 \le \nu \le n)$ is a biholomorphism of Δ and $z_1,\ldots,z_n$ are the coordinates of Δ^n. Because of this assumption on G, the holomorphic tangent bundle of T_N of N is a direct sum of n holomorphic line bundles $L_1 \oplus\cdots\oplus L_n$, with each L_i locally corresponding to a factor of Δ^n. Expressed locally in terms of the coordinates of the factors of Δ^n, the curvature tensor $R_{\alpha\bar\beta\gamma\bar\delta}$ of N is zero unless $\alpha = \beta = \gamma = \delta$; and in that case $R_{\alpha\bar\alpha\alpha\bar\alpha}$ is positive (again we have positivity instead of negativity because of the chosen sign convention). From now on the only local coordinates of N we shall use will be those arising from the coordinates of the factors of Δ^n.

Since

$$R_{\alpha\bar\beta\gamma\bar\delta}\bar\partial f^\alpha\wedge\partial \overline{f^\beta}\wedge\partial f^\gamma\wedge\partial\overline{f^\delta} = \Sigma_\alpha R_{\alpha\bar\alpha\alpha\bar\alpha}\bar\partial f^\alpha\wedge\partial\overline{f^\beta}\wedge\partial f^\gamma\wedge\partial\overline{f^\delta}$$

with $R_{\alpha\bar\alpha\alpha\bar\alpha} > 0$, it follows from the Bochner type argument in (1.2) that both $D\bar\partial f$ and $\partial f^\alpha\wedge\partial f^\alpha$ vanish identically on M for $1 \le \alpha \le n$.

(2.2) <u>Lemma</u>. The line bundle f^*L_α $(1 \le \alpha \le n)$ is a holomorphic line bundle over M when it is given the following holomorphic structure: A smooth local section s of f^*L_α is holomorphic if and only if $\bar D s$ defined by $\bar D s = \bar\partial s + \Gamma^\alpha_{\alpha\alpha} s\bar\partial f$ is identically zero.

<u>Proof</u>. The Frobenius integrability condition of this holomorphic structure is the following. For an arbitrary point x of M choose a local normal coordinate system at $f(x)$ so that each coordinate is a local coordinate of Δ. Then, as usual, the integrability condition is that $\bar\partial_\lambda\bar D_\mu s$ should be symmetric in λ and μ at x. At x,

$$\begin{aligned}\bar\partial_\lambda\bar D_\mu s &= \bar\partial_\lambda(\bar\partial_\mu s + \Gamma^\alpha_{\alpha\alpha} s\bar\partial_\mu f^\alpha)\\ &= \bar\partial_\lambda\bar\partial_\mu s + \Sigma_\beta(\bar\partial_\beta\Gamma^\alpha_{\alpha\alpha})s\ \overline{\partial_\lambda f^\beta}\ \bar\partial_\mu f^\alpha\\ &= \bar\partial_\lambda\bar\partial_\mu s + R_{\alpha\bar\alpha\alpha\bar\alpha}s\ \overline{\partial_\lambda f^\alpha}\ \bar\partial_\mu f^\alpha.\end{aligned}$$

It follows from the vanishing of $\partial f^\alpha\wedge\partial f^\alpha$ on M that $\overline{\partial_\lambda f^\alpha}\ \bar\partial_\mu f^\alpha$ is symmetric in λ and μ and hence that $\bar\partial_\lambda\bar D_\mu s$ is symmetric in λ and μ at x. Q.E.D.

(2.3) <u>Lemma</u>. The form ∂f^α is a holomorphic section of the homomorphic vector bundle $(f^*L_\alpha)\otimes T^*_M$ $(1 \le \alpha \le n)$, where T^*_M is the holomorphic cotangent bundle of M.

Proof. Since $\bar{D}_\mu \partial_\lambda f^\alpha = D_\lambda \bar{\partial}_\mu f^\alpha$ from the definition of covariant differentiation, it follows from $D\bar{\partial}f = 0$ that $\bar{D}\partial f^\alpha = 0$ which means that ∂f is a holomorphic section of $(f^* L_\alpha) \otimes T^*_M$. Q.E.D.

(2.4) Associated to N there is a complex manifold $\bar{N}$ which is anti-biholomorphic to N defined by quotienting the polydisc Δ^n by the conjugates of the group G. By considering the harmonic map from M to $\bar{N}$ which is the composite of f and the antibiholomorphism between N and $\bar{N}$, we conclude that $f^* \overline{L_\alpha}$ $(1 \le \alpha \le n)$ is a holomorphic line bundle over M when it is given the following holomorphic structure: A smooth local section $\bar{s}$ of $f^* \overline{L_\alpha}$ is holomorphic if and only if $\bar{D}s$ defined by $\bar{D}s = \bar{\partial}s + \Gamma^\alpha_{\alpha\alpha} s \bar{\partial} f$ is identically zero. Moreover, $\bar{\partial} f^\alpha$ is a holomorphic section of $(f^* \overline{L_\alpha}) \otimes T^*_M$ $(1 \le \alpha \le n)$. Here $\overline{L_\alpha}$ is the line bundle whose transition functions are the complex conjugate of those of L_α. Instead of arguing with $\bar{N}$ we could get the same conclusions by modifying the arguments of Lemmas (2.2) and (2.3) in the obvious way.

§3. Complex Structure of the Space of Leaves of Certain Foliations.

(3.1) Suppose M is a compact Kähler manifold and L is a holomorphic line bundle over M. Let ξ be a holomorphic section of $L \otimes T^*_M$ over M and let Z be the zero-set of ξ. Assume the following conditions:

(i) Z is of complex codimension at least two in M.

(ii) Every point of M admits an open neighborhood U in M on which there is a local trivialization of L so that, when ξ is regarded as a holomorphic 1-form via this trivialization, one has $\xi \wedge d\xi \equiv 0$.

Note that if $\xi \wedge d\xi \equiv 0$ with respect to some local trivialization of L then $\xi \wedge d\xi \equiv 0$ for all local trivialization of L. On $M - Z$ the condition that $\xi \wedge d\xi \equiv 0$ for some local trivialization of L is the same as the Frobenius condition for the integrability of the distribution defined by the (well-defined) kernel of ξ. Hence on $M - Z$ we have a holomorphic foliation of codimension one defined by the kernel of ξ, henceafter to be called the ξ-foliation.

(3.2) We want to prove that each leaf of the ξ-foliation in $M - Z$ is closed in $M - Z$ (and therefore each has a complex-analytic closure in M by the theorem of Remmert-Stein [24]) and that the set of all leaves of foliation in $M - Z$ (with some leaves suitably grouped together) can be given a holomorphic structure to become a compact complex space. The reason that some leaves of the ξ-foliation in $M - Z$ have to be grouped together to form a single point in the complex space has to do with reducibility and limit behavior. Namely, the limit of a sequence of complex-analytic hypersurfaces which are the topological closures in M of single leaves of the ξ-foliation in $M - Z$ may be a reducible complex-analytic hypersurface in M which is the topological closure of the union of several leaves of the ξ-foliation in $M - Z$.

To do the indicated proof, the first difficulty is to show that every leaf of the ξ-foliation is closed in $M - Z$. Even after we know this, there is still the second difficulty of knowing beforehand which leaves of the ξ-foliation in $M - Z$ should be combined to form a single point in order to make the leaves form a complex space. This is actually equivalent to knowing beforehand when the regular part of the topological closure in M of the union of several leaves of the ξ-foliation in $M - Z$ will be connected. To overcome these difficulties, we shall use the cycle space of all complex-analytic hypersurfaces of M (see Barket [1] and cf. Douady [4]).

We now assume given a smooth family of disjoint connected nonsingular complex-analytic hypersurfaces A_j $(j \in J)$ of M parametrized by a connected real smooth manifold J of real dimension two such that the set $G = \bigcup_{j \in J} A_j$ is nonempty and open and for every j and at every point of $A_j - Z$ the tangent space of A_j is the kernel of ξ. We choose some $j_0 \in J$ and consider the branch χ of this cycle space that contains A_{j_0} as a point. Since M is compact Kähler, it follows that χ is compact (cf. Fujiki [8] and Lieberman [14]). Let $\mathcal{V}$ be the set of all points P of χ such that the complex-analytic hypersurface A_P corresponding to P has the property that the tangent space of A_P at every point of $A_P - Z$ is the kernel of ξ. It is clear that $\mathcal{V}$ contains A_j as point for every $j \in J$. We are going to show that we can use $\mathcal{V}$ to give us the complex structure on the space of leaves of the ξ-foliation.

(3.3) <u>Lemma.</u> $\mathcal{V}$ is a complex-analytic subset of χ.

<u>Proof</u>. Take an element P_0 of χ and let A_{P_0} be the complex-ana-

lytic hypersurface of M corresponding to P_0. We can cover A_{P_0} by a finite number of open subsets U_ν $(1 \le \nu \le k)$ with the following properties:

(i) Each U_ν is of the form $\{|z_1^{(\nu)}| < 1, \ldots, |z_n^{(\nu)}| < 1\}$ for some coordinate chart $z_1^{(\nu)}, \ldots, z_n^{(\nu)}$ of M and U_ν is a relatively compact open subset of that chart.

(ii) L is holomorphically trivial over each U_ν.

(iii) For some $0 < r < 1$, the union of $U_\nu' := \{|z_1^{(\nu)}| < r, \ldots, |z_n^{(\nu)}| < r\}$ for $1 \le \nu \le k$ covers A_{P_0} also.

(iv) $\{|z_1^{(\nu)}| \le 1, \ldots, |z_{n-1}^{(\nu)}| \le 1,\ r \le |z_n^{(\nu)}| \le 1\}$ is disjoint from A_{P_0} and from Z.

Let π_ν be the projection defined with respect to the coordinate chart $(z_1^{(\nu)}, \ldots, z_n^{(\nu)})$ by

$$(z_1^{(\nu)}, \ldots, z_n^{(\nu)}) \to (z_1^{(\nu)}, \ldots, z_{n-1}^{(\nu)}).$$

Then π_ν makes $A_{P_0} \cap U_\nu$ an analytic cover of λ_ν sheets over the $(n-1)$-disc $W_\nu := \pi_\nu(A_{P_0} \cap U_\nu)$. Let E_ν be the complex-analytic hypersurface of W_ν which is the branching locus of the analytic cover $\pi_\nu : A_{P_0} \cap U_\nu \to W_\nu$.

Take a relatively compact non-empty coordinate ball B_ν in $W_\nu - E_\nu - \pi_\nu(Z \cap U_\nu)$. Note that $\pi_\nu(Z \cap U_\nu)$ is a closed subset of W_ν because $\{|z_1^{(\nu)}| \le 1, \ldots, |z_{n-1}^{(\nu)}| \le 1,\ r \le |z_n^{(\nu)}| < 1\}$ is disjoint from Z.

Because of the way the complex structure of X is defined, there exists an open neighborhood D of P_0 in X with the following property: There exist λ_ν holomorphic functions $f_i^{(\nu)}(x,t)$ on $B_\nu \times D$ (with $x \in B_\nu$ and $t \in D$) such that, if A_t is the complex-analytic hypersurface of M corresponding to t, then $A_t \cap U_\nu \cap \pi_\nu^{-1}(B_\nu)$ is given by

$$\bigcup_{i=1}^{\lambda_\nu} \{Q \in U_\nu \cap \pi_\nu^{-1}(B_\nu) \,|\, z_n^{(\nu)}(Q) = f_i^{(\nu)}(\pi_\nu(Q), t)\}.$$

Moreover, $|f_i^{(\nu)}| < 1$ on $B_\nu \times D$ for $1 \le i \le \lambda_\nu$; and at every point of $B_\nu \times D$, $f_i^{(\nu)}$ and $f_j^{(\nu)}$ assume different values if $i \ne j$.

Since L is holomorphically trivial over each U_ν, we can regard $\xi|U_\nu$ as a holomorphic 1-form on U_ν by using a holomorphic trivialization of $L|U_\nu$. For $t \in D$ and $x_\nu \in B_\nu$ and for every type $(1,0)$ tangent vector Y of B_ν at x, let $F_{i,x_\nu,Y}^{(\nu)}(t)$ be the value of the holomorphic 1-form $\xi|U_\nu$ at the tangent vector that is the image of the tangent vector Y under the holomorphic map from B_ν to U_ν that sends $y_\nu \in B_\nu$ to the point $Q \in U_\nu$ with $\pi_\nu(Q) = y_\nu$ and $z_n^{(\nu)}(Q) = f_i^{(\nu)}(y_\nu,t)$.

It is clear that for $t \in D$ we have $t \in \mathcal{V}$ if and only if $F^{(\nu)}_{i,x_\nu,Y}(t) = 0$ for all $1 \le \nu \le k$, $1 \le i \le \lambda_\nu$, $x \in B_\nu$, and all type $(1,0)$ tangent vectors Y of B_ν at x_ν. Now for fixed ν, i, x_ν and Y, the function $F^{(\nu)}_{i,x_\nu,Y}(t)$ of t is a holomorphic function on D. Thus $\mathcal{V} \cap D$ is simply the common zero-set of the holomorphic functions $F^{(\nu)}_{i,x_\nu,Y}(t)$ $(1 \le \nu \le k,\ 1 \le i \le \lambda_\nu,\ x \in B_\nu,\ Y =$ a type $(1,0)$ tangent vector of B_ν at $x_\nu)$. This concludes the proof of the lemma.

(3.4) Let R be the branch of $\mathcal{V}$ that contains A_{j_0} as a point. We have the following holomorphic maps

$$\begin{array}{ccc} \Omega & \xrightarrow{\ \tau\ } & M \\ {\scriptstyle\sigma}\downarrow & & \\ R & & \end{array}$$

where Ω is the complex-analytic subset of $R \times M$ consisting of all points (r,x) such that x belongs to the complex-analytic hypersurface of M corresponding to the point r of R.

From the definition of Z and $\mathcal{V}$ it is clear that τ maps $\tau^{-1}(M - Z)$ biholomorphically onto $M - Z$. Thus R must be of complex dimension one. We give R the normal complex structure and R becomes a compact Riemann surface.

Since every complex-analytic hypersurface of M that intersects $G \cap Z$ must also intersect G it follows from the assumptions on G and $\{A_j\}_{j\in J}$ that τ maps $\tau^{-1}(G)$ biholomorphically onto G. The open subset $\sigma(\tau^{-1}(G))$ of R consists of all points corresponding to the complex-analytic hypersurfaces A_j, $j \in J$, of M.

Because τ maps $\tau^{-1}((M - Z) \cup G)$ biholomorphically onto $(M - Z) \cup G$, it follows that through σ and τ we can define a holomorphic map $\tilde{g}$ from $(M - Z) \cup G$ to R such that $\sigma = \tilde{g}\tau$ on $\tau^{-1}((M - Z) \cup G)$. The map $\tilde{g}$ maps $M - Z - G$ to $R - \sigma(\tau^{-1}(G))$.

Take a point r_0 of $\sigma(\tau^{-1}(G))$. Since $R - r_0$ is an open Riemann surface and is therefore Stein, we can find a finite number of holomorphic functions F_ν $(1 \le \nu \le \ell)$ on $R - \{r_0\}$ embedding it as a closed submanifold of $\mathbb{C}^\ell$. The holomorphic functions $F_\nu g$ on $M - Z - G$ can be extended holomorphically across $Z - G$, because Z is of complex codimension at least two. Hence $\tilde{g}$ can be extended to a holomorphic map g from M to R such that $\sigma = g\tau$ on Ω. This means that the space of leaves of the ξ-foliation in $M - Z$ can be given the complex structure of the compact Riemann surface R (when certain leaves of the ξ-foliation are combined to form single points).

§4. Harmonic Maps Into Riemann Surfaces.

(4.1) We now go back to the situation of a harmonic map f from a compact Kähler manifold M to a compact quotient N of a polydisc Δ^n, as discussed in §2. We consider the special case $n = 1$ when N is a compact hyperbolic Riemann surface. Since there is only one value for the coordinate index α, we denote f^*L_α, ∂f^α, $\bar{\partial} f^\alpha$,... etc. respectively by f^*L, ∂f, $\bar{\partial} f$. We now assume that neither ∂f nor $\bar{\partial} f$ is identically zero.

The holomorphic section ∂f of $(f^*L) \otimes T_M^*$ is not identically zero. Let V be the divisor of ∂f, i.e. the union of the codimension-one branches of the zero-set of ∂f, with multiplicities counted. Let $[V]$ be the holomorphic line bundle over M associated to the divisor V and let s_V be the canonical holomorphic section of $[V]$ whose divisor is V. Let $F = (f^*L) \otimes [V]^{-1}$ and $\xi = (\partial f)s_V^{-1}$. Then ξ is a well-defined holomorphic section of $F \otimes T_M^*$ over M, and the zero-set Z of ξ is of complex codimension at least two in M.

We now verify that with respect to some (and hence every) local trivialization of F we have $\xi \wedge d\xi = 0$. To get a local trivialization of F is the same as getting a nowhere zero local holomorphic section t of the dual bundle F^* of F. We have to verify that the local holomorphic 1-form $t\xi$ satisfies $(t\xi)\wedge d(t\xi) = 0$. Since $t\xi$ is holomorphic, it follows that $d(t\xi) = \partial(t\xi)$ and it suffices to verify $(t\xi)\wedge\partial(t\xi) = 0$ on $M - V$. On $M - V$ the local section t of F^* is of the form $s_V\eta$ where η is a smooth local $f^*T_N^*$-valued function on M with $\bar{D}\eta = 0$. Thus

$$(t\xi)\wedge\partial(t\xi) = (\eta\partial f)\wedge\partial(\eta\partial f) = \eta\partial f\wedge\partial\eta\wedge\partial f,$$

which vanishes, as required.

(4.2) Since the Frobenius integrability condition is precisely that with respect to some local trivialization of F, the local holomorphic 1-form ξ satisfies $\xi\wedge d\xi = 0$, it follows that in $M - Z$ the kernel of ξ defines a holomorphic foliation of complex codimension one, thereafter to be called the ξ-foliation, where Z is the zero-set of ξ as before.

For each $x \in M - Z$ we have a connected local complex-analytic hypersurface H in $M - Z$ such that H contains x and such that the tangent spaces of H are contained in the kernel of ξ. Thus ∂f is identicaly zero when restricted to the tangent spaces of H.

Assume first that H is not contained in V. Since $\partial f \wedge \partial \bar{f} = 0$ on M, it follows that $\partial \bar{f} = \lambda \partial f$ on $M - V$ for some smooth function λ on $M - V$. As a consequence, $\partial \bar{f}$ (and therefore $\bar{\partial} f$) is also identically zero when restricted to the tangent spaces of $H - V$. So $df = \partial f + \bar{\partial} f$ is identically zero when restricted to the tangent spaces of $H - V$ and f is constant along $H - V$ and hence along H.

If, on the other hand, H is contained entirely in V, then H is the limit of a sequence of connected local complex-analytic hypersurfaces H_ν $(1 \le \nu < \infty)$ in $M - Z$ such that the tangent spaces of each H_ν are contained in the kernel of ξ and such that no H_ν is entirely contained in V. Then the above argument shows that f is constant on each H_ν and therefore is constant also on H. Hence at every point of $M - Z$ the leaf of the ξ-foliation is contained in the fiber of f at that point.

(4.3) We now assume that f is surjective. Let E be the set of points of M where the rank over $\mathbb{R}$ of the differential df of f is less than two and let $E' = f(E)$. Since E is compact, E' is also compact. By Sard's theorem, E' is of measure zero in N. For $y \in N - E'$, $f^{-1}(y)$ is a smooth submanifold of real codimension two in M. Since at every point of $f^{-1}(y) - Z$ the leaf of the ξ-foliation is contained in the fiber of f at that point, it follows that $f^{-1}(y)$ is a nonsingular complex-analytic hypersurface. Also, any connected component of $f^{-1}(y) - Z$ is a leaf of the ξ-foliation.

Pick a point y_0 of $N - E'$ and pick a branch A_{y_0} of $f^{-1}(y)$. We can find an open neighborhood G of A_{y_0} in M and an open neighborhood J of y_0 in $N - E'$ such that f maps G into J and for every $y \in J$, $G \cap f^{-1}(y)$ is a connected submanifold of M.

We now apply the result of §3 with $A_j = G \cap f^{-1}(j)$ for $j \in J$ and conclude that there exist a compact Riemann surface R and a holomorphic map g from M to R such that

(i) for $r \in R$, $g^{-1}(r) - Z$ is a leaf of the ξ-foliation,

(ii) for some nonempty open subset R_0 of R there is a diffeomorphism $\theta:R_0 \to J$ such that $g^{-1}(r) = G \cap f^{-1}(\theta(r))$ for $r \in R_0$.

(4.6) By regarding every connected topological component of a fiber of $g:M \to R$ as a point, we can construct a complex space $\tilde{R}$ with a holomorphic map $\pi:\tilde{R} \to R$ (see e.g. [12]). Since $\pi:\tilde{R} \to R$ is an analytic cover and π maps $\pi^{-1}(R_0)$ biholomorphically onto R_0, it follows that π maps $\tilde{R}$ biholomorphically onto R and every fiber of $g:M \to R$ is connected.

Because at every point of $M - Z$ the leaf of the ξ-foliation is contained in the fiber of f at that point and because every fiber of $g:M \to R$ is connected, we conclude that there exists a continuous map $\varphi:R \to N$ such that $f = \varphi g$.

By the theorem of Eells-Sampson there exists a harmonic map $\psi:R \to N$ in the same homotopy class as φ. Since R is a Riemann surface and g is holomorphic, ψg is a harmonic map from M to N. Because ψg is in the same homotopy class as $\varphi g = f$, we have $\psi g = f$ by Hartman's result on the uniqueness of harmonic maps [9]. It follows from the surjectivity of g that $\psi = \varphi$ and φ is smooth and harmonic.

The Riemann surface R must be hyperbolic. Otherwise R could be given a conformal metric of nonnegative curvature. The map φ would still be harmonic with respect to this metric. But then $\varphi:R \to N$ would have to have real rank < 2, because a harmonic map from a compact Riemann manifold of nonnegative Ricci curvature to a compact Riemannian manifold of negative sectional curvature cannot have real rank ≥ 2 (see e.g. [7]). This would contradict the surjectivity of φ.

By using the theorem of Eells-Sampson on the existence of harmonic maps [7] and the fact that any smooth map from a compact Kähler manifold to a compact Riemann surface which is nonzero on the second homology group must be surjective, we obtain from the above discussion the following theorem.

(4.7) Theorem. Let M be a compact Kähler manifold and N be a compact hyperbolic Riemann surface such that there exists a continuous map h from M to N which is nonzero on the second homology group. Then there exists a holomorphic map g from M to a compact hyperbolic Riemann surface R and a harmonic map φ from R to N such that $h = \varphi g$. As a consequence the lifting of g to the universal covers of M and R is a nonconstant bounded holomorphic function on the universal cover of M.

§5. Strong Rigidity of Irreducible Quotients of Polydiscs.

(5.1) After the strong rigidity of compact quotients of irreducible bounded symmetric domains of complex dimension at least two was proved in [26,27,28] the only case of the complex-analytic analog of Mostow's strong rigidity that was not yet settled was that of irreducible compact quotients of polydiscs of complex dimension at least two. For the case of a compact quotient of an irreducible bounded symmetric domain of complex dimension at least two, the curvature tensor is negative enough to yield the holomorphicity or antiholomorphicity for harmonic maps from a compact Kähler manifold into it with sufficiently high rank at some point. For the case of an irreducible quotient of a polydisc, the curvature tensor fails to have enough negativity because of the vanishing of the holomorphic bisectional curvature of the polydisc in the direction of any pair of vectors tangential to two different component discs. Jost-Yau [10] first constructed holomorphic foliations on a compact Kähler manifold from a harmonic map into a quotient of a polydisc which is neither holomorphic nor antiholomorphic (even after replacing some factors of the polydisc by their complex-conjugates) and satisfies a rank condition. From that they derived the holomorphicity of harmonic maps into compact quotients of polydiscs under some additional conditions on Chern classes but without the irreducibility condition for the quotient [10, Th. 6.1]. The Chern class conditions were needed to handle the situation for the irreducible case when their holomorphic foliations become singular (see [10], Sect. 2, p. 150]). Later Jost-Yau [11] proved the strong rigidity of irreducible compact quotients of a polydisc for the case of complex dimension two. Mok [17] finally proved the general case by using extension results for complex-analytic subvarieties to handle the singularity of the holomorphic foliations and using the result of Borel [2] and Matsushima-Shimura [15] to get the final conclusion from the irreducibility of the compact quotient.

(5.2) <u>Theorem</u> (Jost-Yau [11] and Mok [17]). Suppose that Q is an irreducible compact quotient of an n-disc Δ^n with $n \geq 2$, that M is a compact Kähler manifold, and that f is a harmonic map from M to Q which is a homotopy equivalence. Let $\pi:\tilde{M} \to M$ be the universal cover of M and $F:\tilde{M} \to \Delta^n$ with components $(F^1,\ldots,F^n)$ be the map induced by F. Then for each $1 \le i \le n$, F^i is either holomorphic or antiholomorphic. As a consequence, f is a diffeomorphism.

Here an irreducible quotient means one that cannot be decomposed as a product of two lower-dimensional quotients of polydiscs. By using our result on the holomorphic structure of the pullback of the tangent bundle under certain harmonic maps discussed in §2, we can give a more natural interpretation of the holomorphic foliation of Jost-Yau [11] which automatically takes care of the situation when the holomorphic foliation becomes singular. We present here a more streamlined proof of the theorem stated above by using our interpretation of the holomorphic foliation and getting the existence of a closed leaf by a topological argument. Moreover, in the last step of using the results of Borel [2] and Matsushima-Shimura [15] to obtain the final conclusion, we use an argument that avoids the difficulty that arises because the correspondence between the fundamental group and the group of covering transformations of the universal cover is not unique but is determined only up to an inner isomorphism (see (5.4)).

(5.3) Let $q:\Delta^n \to Q$ be the quotient map. For the proof of Theorem (5.2), by replacing M and Q by finite covers, we can assume without loss of generality that the group G of all covering transformations of $q:\Delta^n \to Q$ is contained in $(\text{Aut }\Delta)^n$, where Aut means the automorphism group. For $1 \le i \le n$ let G_i be the set of all $g_i \in \text{Aut }\Delta$ such that $(g_1,\ldots,g_{i-1},g_i,g_{i+1},\ldots,g_n)$ belongs to G for some $g_1,\ldots,g_{i-1},g_{i+1},\ldots,g_n$ in Aut Δ. Then G is contained in the product of the n groups $G_1,\ldots,G_n$. (The words "contained in" are mistakenly missing in the corresponding statement in the sketch of the proof given in [29]. There it should also be stated that the product is the smallest product that contains G.) The irreducibility of Q will be used to invoke the results of Borel [2] and Matsushima-Shimura [15] to conclude that every orbit of G_i $(1 \le i \le n)$ is dense in Δ.

To prove Theorem (5.2) we have to show that for every $1 \le i \le n$ either ∂F^i or $\bar{\partial} F^i$ vanishes identically on M. Without loss of generality we assume that the assertion fails for $i = 1$ and try to get a contradiction. We use the notations of §2. We know that ∂f^1 (respectively $\bar{\partial} f^1$) is a holomorphic section of the holomorphic line bundle $f^*L_1 \otimes T_M^*$ (respectively $f^*\overline{L_1} \otimes T_M^*$). As in (4.1) we let V (respectively V') be the divisor of M which is the union of the branches of the zero-set of ∂f^1 (respectively $\partial \overline{f^1}$) of complex codimension 1 with multiplicities counted. Let s_V (respectively $s_{V'}$) be the canonical holomorphic section of the line bundle $[V]$ (respectively $[V']$) whose divisor is V (respectively V'). Let ξ (respectively ξ') be the holomorphic section $(\partial f^1)s_V^{-1}$ (respectively $(\partial \overline{f^1})s_{V'}^{-1}$) of the holomorphic line bundle $(f^*L_1) \otimes [V]^{-1} \otimes T_M^*$ (respectively $(f^*\overline{L_1}) \otimes [V']^{-1} \otimes T_M^*$). Let Z (respectively Z') be the zero-set of ξ (respectively ξ'). Then Z and Z' are both of complex codimension at least two. As in (4.1) we know that the kernel of ξ (respectively ξ') defines a holomorphic foliation of complex codimension one in $M - Z$ (respectively $M - Z'$). Since $\partial f^1 \wedge \partial \overline{f^1}$ is identically zero on M, it follows that on $M - (Z \cup Z')$ the holomorphic foliation defined by the kernel of ξ agrees with that defined by the kernel of ξ'. Hence in $M - (Z \cap Z')$ there is a holomorphic foliation $\mathcal{F}$ of complex codimension one defined by the kernel of ξ or the kernel of ξ'.

Since ∂f^1 and $\partial \overline{f^1}$ both vanish on the tangent spaces of leaves of the foliation $\mathcal{F}$, it follows from $df^1 = \partial f^1 + \overline{\partial \overline{f^1}}$ that the locally defined function f^1 is locally constant on every leaf of the foliation $\mathcal{F}$. Because of the existence of the foliation $\mathcal{F}$, every point of $M - (Z \cap Z')$ admits an open neighborhood U in $M - (Z \cap Z')$ with the following property: On U the function f^1 can be globally defined and there exists a holomorphic map h from U to an open subset G of $\mathbb{C}$ with nowhere zero differential such that every connected component of a fiber of h is an open subset of a leaf of the foliation $\mathcal{F}$, and, moreover, there exists a smooth function t on G such that $f^1 = t\, h$. Therefore the zero-set of ∂f^1 (respectively $\partial \overline{f^1}$) in U agrees with the inverse image under h of the zero-set of ∂t (respectively $\partial \overline{t}$). It follows that the regular part in $M - Z$ of every branch of V or V' is a leaf of the foliation $\mathcal{F}$.

Since ∂f^1 and $\overline{\partial f^1}$ are linearly dependent over $\mathbb{C}$ at every point of M, it follows that there exists a unique meromorphic section τ of $(f^*L_1) \otimes (f^*\overline{L_1})^{-1}$ such that $\xi = \tau\xi'$ and the divisor of τ is $V - V'$. The first Chern class $c_1(L_1 \otimes \overline{L}_1^{-1})$ of $L_1 \otimes \overline{L}_1^{-1}$ is a nonzero cohomology class in Q, because it is equal to $2c_1(L_1)$ and is represented by a nonzero multiple of the 2-form on Q coming from the Kähler form of the first factor of Δ^n. Since f is a homotopy equivalence, the holomorphic line bundle $(f^*L_1) \otimes (\overline{f^*L_1})^{-1}$, being topologically the pullback of $L_1 \otimes \overline{L}_1^{-1}$, has a nonzero first Chern class and the divisor of τ cannot be zero. Let C be a branch of the divisor of τ. Since C is either a branch of V or a branch of V', it follows that the regular part of C in $M - Z$ is a leaf of the foliation $\mathcal{F}$. Hence F^1 is constant on every connected component of $\pi^{-1}(C)$.

(5.4) Choose a point x of C and a point $\tilde{x}$ of $\pi^{-1}(C)$. Let $\tilde{C}$ be the connected component of $\pi^{-1}(C)$ that contains $\tilde{x}$. Let Σ be the fundamental group $\pi_1(M,x)$ of M with base point x. Let S be the group of all covering transformations of the topological covering $\pi:\tilde{M} \to M$. Let $y = f(x)$ and choose y as the base point of Q. Let Γ be the fundamental group $\pi_1(Q,y)$ with base point y. Recall that G is the group of all covering transformations of the topological covering $q:\Delta^n \to Q$. Let $\tilde{y} = F(\tilde{x})$.

There is a standard one-one correspondence between Σ and S which depends on the choice of $\tilde{x}$. It is given as follows. For any $s \in S$, $s(\tilde{x})$ belongs to $\pi^{-1}(x)$. Join $\tilde{x}$ to $s(\tilde{x})$ by a path ρ. Then the element σ of Σ defined by the loop $\pi\rho$ corresponds to s. Conversely, if an element σ of Σ is represented by a loop θ in M starting at x, then θ can be lifted up to a path in $\tilde{M}$ starting from $\tilde{x}$ and ending at some point x^* of $\pi^{-1}(x)$. Then the element of S which maps $\tilde{x}$ to x^* corresponds to σ. (Note that if another element $\hat{x}$ in $\pi^{-1}(x)$ is chosen instead of $\tilde{x}$, then the correspondence between Σ and S is changed by an inner automorphism of S defined by the element of S which maps $\tilde{x}$ to $\hat{x}$.) Likewise there is a one-one correspondence between Γ and G which depends on $\tilde{y}$.

Let $I:C \hookrightarrow M$ be the inclusion map and let $\Gamma^0 = f_* i_*(\pi_1(C,x))$ and let G^0 be the subset of G corresponding to Γ^0 under the correspondence between Γ and G which depends on $\tilde{y}$. Write $\tilde{y} = (\tilde{y}^1,\ldots,\tilde{y}^n)$. Now F^1 is constant on $\tilde{C}$. Take an arbitrary element γ^0 of Γ^0. It is represented by a loop $f\theta$ in C, where θ is a loop starting at x. We lift the loop θ to a path $\tilde{\theta}$ in $\tilde{M}$ starting at $\tilde{x}$ and ending at some x^*. Since the loop θ is inside C, the path $\tilde{\theta}$, being continuous, must be inside $\tilde{C}$. So $F^1(x^*) = F^1(\tilde{x}) = \tilde{y}^1$. The element g^0 of G^0 which corresponds to γ^0 sends x to x^*. Since we can write $g^0 = (g_1^0,\ldots,g_n^0)$, it follows that g_1^0

fixes $\tilde{y}^1$ and g^0 fixes every point of $\tilde{y}^1 \times \Delta^{n-1}$. Hence the map $F:\tilde{C} \to \tilde{y}^1 \times \Delta^{n-1}/G^0$ descends to a map $\varphi: C \to \tilde{y}^1 \times \Delta^{n-1}/G^0$. The map $f: C \to f(C)$ factors through φ and $\tilde{y}^1 \times \Delta^{n-1}/G^0 \to \tilde{y}^1 \times \Delta^{n-1}/G' \hookrightarrow Q$, where G' consists of all elements of G which fix $\tilde{y}^1 \times \Delta^{n-1}$.

(Note that if there is another connected component $\tilde{C}_1$ of $\pi^{-1}(C)$ so that the point $F^1(\tilde{C}_1)$ is different from the point $F^1(\tilde{C})$, it does not immediately follow that the first component of every element of G^0 as an element of $\operatorname{Aut}\Delta$ fixes the point $F^1(\tilde{C}_1)$, because G^0 depends on the point $\tilde{y}$. We can only conclude that there exists a subgroup $\hat{G}$ of G conjugate to G^0 such that the first component of every element of $\hat{G}$ fixes the point $F^1(\tilde{C}_1)$.)

We have the following commutative diagram

$$\begin{array}{ccccc} H_{2n-2}(C,\mathbb{Z}) & \xrightarrow{i_*} & H_{2n-2}(M,Z) & \xrightarrow{f_*} & H_{2n-2}(Q,\mathbb{Z}) \\ & {\scriptstyle\varphi_*}\searrow & & \nearrow & \\ & & H_{2n-2}(\tilde{y}^1 \times \Delta^{n-1}/G^0,\mathbb{Z}) & & \end{array}$$

Since M is Kahler, i_* is injective. By assumption f_* is an isomorphism. Hence φ_* is injective. Since $\tilde{y}^1 \times \Delta^{n-1}/G^0$ is a complex manifold of complex dimension $n-1$, φ_* is injective only if $\tilde{y}^1 \times \Delta^{n-1}$ is compact and φ is surjective. This implies that $F:\tilde{C} \to \tilde{y}^1 \times \Delta^{n-1}$ is surjective. Since $\tilde{x}$ is an arbitrarily chosen point of $\pi^{-1}(C)$, it follows that whenever $F(\pi^{-1}(C))$ contains a point $\tilde{y}$ of Δ^n, it contains the whole fiber $pr_1^{-1}(pr_1(\tilde{y}))$, where $pr_1:\Delta^n \to \Delta$ is the projection onto the first factor.

Since f maps Σ isomorphically onto Γ, it follows that $F(\pi^{-1}(C)) = q^{-1}(f(C))$ and $F(\pi^{-1}(C))$ is closed and invariant under G. Because $F(\pi^{-1}(C))$ is a union of fibers of the map $pr_1:\Delta^n \to \Delta$, we conclude that $pr_1(F(\pi^{-1}(C)) = F^1(\pi^{-1}(C))$ is a closed subset of Δ and is invariant under the action of G_1. This contradicts the results of Borel [2] and Matsushima-Shimura [15], which has as a consequence the density of any orbit in Δ of the subgroup G_1 of $\operatorname{Aut}\Delta$.

§6. Holomorphicity of Harmonic Maps of Finite Energy.

(6.1) The result in this section is the joint work with S.-T. Yau.

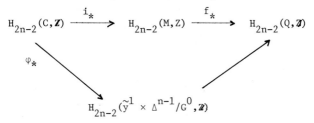

Mostow's strong rigidity theorem holds not only in the compact case but also in the case of finite volume. For the complex-analytic analog of strong rigidity, in the biholomorphism result the curvature condition should be imposed only on one of the two Kähler manifolds. If one imposes only the condition of finite volume on the other Kähler manifold, it is clearly too weak a condition because without any curvature condition one can in many cases rather easily change the complete Kahler metric to make the volume of the manifold finite. However, one can obtain results by considering harmonic maps of finite global energy. Recall that the global energy means the global L^2 norm over the domain manifold of the differential of the map. For the proof of the complex-analytic analog of strong rigidity the compactness condition is used so that Stokes' theorem can be applied without worrying about any boundary terms. In the noncompact case boundary terms would occur. However, if one assumes that the harmonic map has finite global energy, the contribution from the boundary can be shown to have zero limit when one applies Stokes' theorem by using a suitable sequence of cut-off functions. The key step is the following lemma.

(6.2) Lemma. Let M and N be Riemannian manifolds such that M is complete, the Riemannian sectional curvature of N is nonpositive, and the Ricci curvature of M is bounded from below by a negative number $-k$. Let $f:M \to N$ be a harmonic map with finite global energy. Then the global L^2 norm over M of the covariant derivative ∇df of the differential df of f is finite.

Proof. Let ∇ denote the covariant differential operator for tensors and Δ be the Laplace-Beltrami operator for functions. By the divergence of a 1-form we mean the divergence of the vector field associated to the 1-form and we use the abbreviation div for divergence. Unless specified to the contrary, integration is over all of M.

Take an arbitrary compact subset K of M. By the completeness of M there exists a smooth function $0 \le \varphi \le 1$ on M such that φ is identically 1 on K and the pointwise norm of $\nabla\varphi$ is bounded by a constant A on M. Let e be the pointwise energy function of f which is defined as the pointwise L^2 norm of the differential df of f. Since the Riemannian sectional curvature of N is nonpositive and the Ricci curvature of M is bounded from below by $-k$, it follows from straightforward direct computation that $\Delta e \ge |\nabla df|^2 - ke$. From the divergence theorem,

$$0 = \int \mathrm{div}(\varphi^2 \nabla e) = 2\int \varphi \nabla\varphi \cdot \nabla e + \int \varphi^2 \Delta e$$

$$\geq 2\int \varphi \nabla\varphi \cdot \nabla e + \int \varphi^2 |\nabla df|^2 - \int k\varphi^2 e.$$

$$\left|\int \varphi \nabla \varphi \cdot \nabla e\right| \leq \left(\int e\right)^{1/2} \left(\int \varphi^2 |\nabla\varphi|^2 \frac{|\nabla e|^2}{e}\right)^{1/2}$$

$$\leq 8A^2 \int e + \frac{1}{16A^2} \int \varphi^2 |\nabla\varphi|^2 \frac{|\nabla e|^2}{e}$$

$$\leq 8A^2 \int e + \frac{1}{4}\int \varphi^2 |df|^2,$$

where for the last inequality we have used $|\nabla e|^2 \leq 4|\nabla df|^2 e$ from direct covariant differentiation of $e = |df|^2$ and the Schwarz inequality. So

$$\int \varphi^2 |\nabla df|^2 \leq 16A^2 \int e + \frac{1}{2}\int \varphi^2 |\nabla df|^2 + \int k\varphi^2 e.$$

$$\int \varphi^2 |\nabla df|^2 \leq 32A^2 \int e + 2k\int \varphi^2 e.$$

Since K is an arbitrary compact subset of M, it follows that

$$\int |\nabla df|^2 \leq (32A^2 + 2k)\int e.$$

Thus the global L^2 norm of ∇df over M is bounded. Q.E.D.

For the sake of simplicity we state the following theorem only for the case of strongly negative curvature. The method clearly works to give results for the same general situation as in the compact case.

(6.3) Theorem. Suppose M and N are Kähler manifolds such that M is complete, the curvature of N is strongly negative, and the Ricci curvature of M is bounded from below. Let $f: M \to N$ be a harmonic map with finite global energy. If the rank of f over $\mathbb{R}$ is at least 3 at some point, then f is either holomorphic or antiholomorphic.

Proof. Let $h_{\alpha\bar\beta}$ be the Kähler metric of N and ω be the Kähler form of M. Let m be the complex dimension of M. Since M is complete, there exist a sequence of smooth functions $0 \leq \varphi_i \leq 1$ $(1 \leq i < \infty)$ on M with compact support such that φ_{i+1} is identically 1 on the support of φ_i and $\nabla\varphi_i$ is bounded by $1/i$ on M. By Stokes' theorem the left-hand side of the following equation is zero.

$$\int d(\varphi_\nu \bar\partial (h_{\alpha\bar\beta} \bar\partial f^\alpha \wedge \partial \overline{f^\beta} \wedge \omega^{m-2})) = \int \partial\varphi_\nu \wedge h_{\alpha\bar\beta} \bar\partial f^\beta \wedge \bar{D} \overline{\partial f^\beta} \wedge \omega^{m-2}$$

$$+ \int \varphi_\nu \partial\bar\partial (h_{\alpha\bar\beta} \bar\partial f^\alpha \wedge \partial \overline{f^\beta} \wedge \omega^{m-2}),$$

where $\bar{D}$ means covariant differentiation as in (1.2). By using the Schwarz inequality once, we conclude that the first term on the right-hand side approaches zero as i goes to infinity, because $\nabla\varphi_i$ is bounded by $1/i$ on M and because of the finiteness of the global energy of f and Lemma (6.2). By using the second term on the right-hand side one can argue as in the compact case to get the conclusion of the theorem.

We now derive from the above theorem a strong rigidity result for the noncompact case. First we obtain the noncompact case of the existence of harmonic maps with finite energy.

(6.4) Theorem. Let M and N be complete Riemannian manifolds. Suppose the Riemannian sectional curvature of N is bounded from above by a negative number and $f: M \to N$ is a smooth map with finite global energy such that the image under f of the fundamental group of M is not a nilpotent group. Then f is homotopic to a harmonic map with finite global energy.

Proof. Let M_i $(1 \leq i < \infty)$ be an increasing sequence of relatively compact domains in M with smooth boundary whose union is M. According to the theorem of R. Schoen [25, p. 115, Chap. IV, Sect. 3] there exists a harmonic map h_i from the closure $\overline{M_i}$ of M_i to N which is homotopic to $f|\overline{M_i}$ and agrees with f on the boundary of M_i. By Schoen's procedure of constructing h_i we know that on each fixed connected compact subset D of M the family of maps h_i $(1 \leq i < \infty)$ is equicontinuous with Lipschitz estimates. To prove that there is a subsequence h_i which converges, it suffices to prove that for some point x of D the set $\{h_i(x)\}$ is bounded in N for some subsequence of h_i. Suppose the contrary. Then $h_i(D)$ converges to the boundary of N and hence lies eventually in some end of N. In this situation the image of the fundamental group of D under f must be nilpotent, which is a contradiction when D is sufficiently large (see e.g. [6] or [30]). Hence h_i converges to a harmonic map from M to N with finite global energy which is homotopic to f, because the global energy of h_i is dominated by that of f. Q.E.D.

Though the method works in more general situations, for the sake of simplicity we state the strong rigidity result for the noncompact finite-volume case only in the following simple situation. For needed information about the geometry of a complete negatively curved Riemannian manifold of finite volume see e.g. [6] or [30].

(6.5) Theorem. Let M and N be complete Kähler manifolds of complex dimension ≥ 2 with finite volume whose Riemannian sectional curvatures are bounded between two negative numbers $-a$ and $-b$ with $a > b$. Assume that the curvature of N is strongly negative. If M and N are of the same homotopy type, then they are either biholomorphic or antibiholomorphic.

Proof. Without loss of generality we can assume that M is noncompact, because the compact case is already covered by the methods of [26]. Because of the curvature and finite-volume assumptions M (respectively N) is the union of a compact subset K_M (respectively K_N) and a finite number of ends M_i (respectively N_i). Each end M_i (respectively N_i) is diffeomorphic to the product of a compact smooth manifold S_i (respectively T_i) and a half-line $[0,\infty)$ under a map $\varphi_i:S_i \times [0,\infty) \to M_i$ (respectively $\psi_i:T_i \times [0,\infty) \to N_i$). For x in S_i (respectively T_i) the diffeomorphism φ_i (respectively ψ_i) maps $x \times [0,\infty)$ to a geodesic ray in M (respectively N) whose arc-length is measured by the coordinate of $[0,\infty)$. For $0 \leq t < \infty$ the diffeomorphism φ_i (respectively ψ_i) maps $T_i \times t$ (respectively $S_i \times t$) to a submanifold perpendicular to those geodesic rays. The fundamental group of M_i (respectively N_i) is characterized by being a maximal noncyclic nilpotent subgroup of the fundamental group of M (respectively N).

After enlarging K_N and shrinking T_i if necessary, we can find a smooth map g from M to N mapping each end M_i of M into an end N_i of N. We shall use the same indexing for the corresponding ends of M and N. Moreover, we can change the homotopy equivalence g so that the map from $S_i \times [0,\infty)$ to $T_i \times [0,\infty)$ which corresponds to it sends (x,t) to $(g_i(y), ct + t_0)$ for some smooth map g_i from S_i to T_i and some c greater than a/b and some positive number t_0. Then the global energy of g is finite. Since g is a homotopy equivalence, the fundamental group of M cannot be mapped by g to a nilpotent subgroup of the fundamental group of N (see e.g. [5] and [16]).

By Theorem (6.4) we can construct a harmonic map f from M to N which is a homotopy equivalence. We are going to show that the real rank of f is at least $2n - 1 \geq 3$. When M has more than one end, one can see this very easily, because for any fixed i the homology class defined by $\varphi_i(S_i \times t)$ is nonzero for any sufficiently large t and must be mapped to a nonzero homology class in N by f.

For the case when M has only one end we argue as follows. Let M^* (respectively N^*) be the universal cover of M (respectively N). Let M_1 (respectively N_1) be the only end of M (respectively N). The fundamental group of M_1 (respectively N_1) is naturally a subgroup of the fundamental group of M (respectively N) and we denote that subgroup by G_1 (respectively H_1). Let $\hat{M}$ (respectively $\hat{N}$) be the quotient of M^* (respectively N^*) with respect to the group G_1 (respectively H_1). We lift the map $f:M \to N$ to a map $\hat{f}:\hat{M} \to \hat{N}$. For any sufficiently large t the real submanifold $\varphi_1(S_1 \times t)$ of M can be lifted to some real submanifold R_t of $\hat{M}$. The homology class in $\hat{M}$ defined by R_t is nonzero for any sufficiently large t and must be mapped to a nonzero homology class in $\hat{N}$ by $\hat{f}$. Hence the real rank of $\hat{f}$ must be at least $2n-1$ at some point of $\hat{M}$ and the real rank of f is at least $2n-1 \geq 3$ at some point of M. (This argument of proving the real rank of $f \geq 2n-1$ in the case of a single end was orally communicated to us by David Kazhdan.) By Theorem (6.3) the harmonic map f is either holomorphic or antiholomorphic. By replacing N by its complex conjugate if necessary, we can assume without loss of generality that f is holomorphic.

By the result of [30], M (respectively N) can be compactified into a compact complex space $\tilde{M}$ (respectively $\tilde{N}$) by adding a finite number of points u_i (respectively v_i), one for each end. Cover $\tilde{N}$ by a finite number of Stein open subsets U_j of $\tilde{N}$. There exists a positive number ε such that any set of diameter $\leq \varepsilon$ in $\tilde{N}$ is contained in some U_j. For each k, the diameter of $\varphi_k(S_k \times t)$ goes to zero as t goes to infinity. By the Schwarz lemma for complete Kähler manifolds proved by Yau [31], the diameter of $f(\varphi_k(S_k \times t))$ goes to zero as t goes to infinity. Hence, for t sufficiently large, $f(\varphi_k(S_k \times t))$ is contained in U_j. Since $\varphi_k(S_k \times t)$ is the strongly pseudoconvex boundary of a Stein open neighborhood of u_k in $\tilde{M}$ and k is arbitrary, the holomorphic map f can be extended to a holomorphic map $\tilde{f}$ from $\tilde{M}$ to $\tilde{N}$. Because f maps the fundamental group of $\varphi_k(S_k \times t)$ isomorphically onto the subgroup of the fundamental group of N corresponding to $\psi_k(T_k \times t)$, it follows that $\tilde{f}(u_k)$ cannot be a point of N; otherwise $f(\varphi_k(S_k \times t))$ would be contained in a contractible subset of N for each t sufficiently large. We conclude that f is proper. Because the real rank of f is at least $2n-1$, we know that f is surjective. Since f does not map the homology class represented by any positive-dimensional subvariety of M to zero, it follows that f is an analytic cover. By considering the map induced by f on the fundamental group of M_i, we conclude that the analytic cover $f: M \to N$ has only one sheet and f is a biholomorphism. Q.E.D.

References

1. D. Barlet, Espace analytique réduit des cycles analytiques complexes compacts d'un espace analytique complexe de dimension finie, Séminaire Norquet 1974-75, Springer Lecture Notes in Mathematics 482 (1975), 1-158.

2. A. Borel, Density properties for central subgroups of semi-simple groups without compact components, Ann. of Math. 72 (1960), 179-188.

3. P. Deligne and G.D. Mostow, Hypergeometric functions and non-arithmetic monodromy groups.

4. A. Douady, Le problème des modules pour les sous-espaces analytiques d'un espace analytique donné, Ann. Inst. Fourier (Grenoble) 16 (1966), 1-95.

5. P. Eberlein, Some properties of the fundamental group of a Fuchsian manifold, Invent. Math. 19 (1973), 5-13.

6. P. Eberlein, Lattices in spaces of nonpositive curvature, Ann. of Math. 111 (1980), 436-376.

7. J. Eells and J.H. Sampson, Harmonic mappings of Riemannian manifolds, Amer. J. Math. 86 (1964), 109-160.

8. A. Fujiki, Closedness of the Douady spaces of compact Kähler spaces, Publ. Res. Inst. Math. Sci. 14 (1978/79), 1-52.

9. P. Hartman, On homotopic harmonic maps, Canadian J. Math. 19 (1967), 373-387.

10. J. Jost and S.-T. Yau, Harmonic mappings and Kähler manifolds, Math. Ann. 262 (1983), 145-166.

11. J. Jost and S.-T. Yau, A strong rigidity theorem for a certain class of compact analytic surfaces, Math. Ann. 271 (1985), 143-152.

12. B. Kaup, Über offene analytische Äquivalenzrelationen auf komplexen Raümen, Math. Ann. 183 (1969), 6-16.

13. P. Lelong, Fonctions entières (n variables) et fonctions plurisousharmoniques d'order fini dans $\mathbb{C}^n$, J. Analyse Math. 12 (1964), 365-407.

14. D. Liberman, Compactness of the Chow scheme: applications to automorphisms and deformations of Kähler manifolds, Seminaire

Norquet 1975-77, Springer Lecture Notes 670 (1978), 140-186.

15. Y. Matsushima and G. Shimura, On the cohomology groups attached to certain vector-valued differential forms on products of the upper half plane, Ann. of Math. 78 (1963), 417-449.

16. J. Milnor, A note on curvature and fundamental group, J. Diff. Geom. 2 (1968), 1-7.

17. N. Mok, The holomorphic or anti-holomorphic character of harmonic maps into irreducible compact quotients of polydiscs, Math. Ann. 272 (1985), 197-216.

18. G.D. Mostow, Strong rigidity of locally symmetric spaces, Ann. of Math. Studies 78, Princeton University Press, Princeton, 1973.

19. G.D. Mostow, Existence of a nonarithmetic lattice in SU(2,1) (research announcement), Proc. Natl. Acad. Sci. USA 75 (1978), 3209-3033.

20. G.D. Mostow, On a remarkable class of polyhedra in complex hyperbolic space, Pacific J. Math 86 (1980), 171-276.

21. G.D. Mostow and Y.-T. Siu, A compact Kähler surface of negative curvature not covered by the ball, Ann. of Math. 112 (1980), 321-360.

22. S. Nakano, On complex analytic vector bundles, J. Math. Soc. Japan 7 (1955), 1-12.

23. E. Picard, Sur les fonctions hyperfuchsiennes provenant des séries hypergéometriques de deux variables, Ann. E.N.S. 2 (1885), 357-384.

24. R. Remmert and K. Stein, Über die wesentlichen Singularitäten analytischer Mengen, Math. Ann. 126 (1953), 263-306.

25. R.M. Schoen, Existence and regularity theorems for some geometric variational problems, Ph.D. thesis, Stanford University 1977.

26. Y.-T. Siu, The complex-analyticity of harmonic maps and the strong rigidity of compact Kähler manifolds, Ann. of Math. 112 (1980), 73-111.

27. Y.-T. Siu, Strong rigidity of compact quotients of exceptional bounded symmetric domains, Duke Math. J. 48 (1981), 857-871.

28. Y.-T. Siu, Complex-analyticity of harmonic maps, vanishing and Lefschetz theorems, J. Diff. Geom. 17 (1982), 55-138.

29. Y.-T. Siu, Some recent results in complex manifold theory related to vanishing theorems for the semipositive case, Proceedings of 1984 Bonn Arbeitstagung, Springer Lecture Notes in Mathematics 1111 (1985), 169-192.

30. Y.-T. Siu and S.-T. Yau, Compactification of negatively curved complete Kähler manifolds of finite volume, Ann. of Math. Studies 102 (1982), 363-380.

31. S.-T. Yau, A general Schwarz lemma for Kähler manifolds, Amer. J. Math. 100 (1978), 197-203.

32. J.-q. Zhong, The degree of strong nondegeneracy of the bisectional curvature of exceptional bounded symmetric domains, Proc. Intern. Conf. Several Complex Variables, Hangzhou, ed. Kohn, Lu, Remmert & Siu, Birkhäuser, Boston, 1984, pp. 127-139.

DEPARTMENT OF MATHEMATICS
HARVARD UNIVERSITY
CAMBRIDGE, MASSACHUSETTS 02138

LATTICES IN SEMISIMPLE GROUPS AND INVARIANT

GEOMETRIC STRUCTURES ON COMPACT MANIFOLDS

Robert J. Zimmer

TABLE OF CONTENTS

LATTICES IN SEMISIMPLE GROUPS AND INVARIANT GEOMETRIC STRUCTURES ON COMPACT MANIFOLDS[1]

by Robert J. Zimmer

1. Introduction of the problem and statement of main conjectures and results.

The aim of this paper is to describe a geometrization of the Mostow-Margulis theory of rigidity and representations of discrete subgroups of semisimple groups. More precisely let H be a connected semisimple Lie group and $\Gamma \subset H$ a lattice subgroup. Let G be another Lie group. The general problem considered by Mostow and Margulis was to study the homomorphisms $\pi:\Gamma \to G$. The Mostow rigidity theorem of course deals with the case in which G is semisimple and $\pi(\Gamma)$ is a lattice in G, and the Margulis superrigidity theorem deals with the more general case in which $\pi(\Gamma)$ is merely assumed to be Zariski dense in G. While we shall recall the precise results later, we simply remark here that the ultimate conclusion is that one can essentially understand all such homomorphisms. Roughly speaking, π either extends to a smooth homomorphism of H or $\pi(\Gamma)$ has compact closure (in which case one also has information on the closure), or is a combination of these cases. A geometric generalization of the notion of a homomorphism $\Gamma \to G$ is of course the notion of an action of Γ by automorphisms of a principal G-bundle. Thus, by the geometrization of the Mostow-Margulis theme, we mean the following:

Problem 1. With H,Γ,G as above, understand the actions of Γ by automorphisms of a principal G-bundle $P \to M$, where M is some (say compact) Γ-space.

This of course is an extremely broad question. Letting P be the principal $GL(n,\mathbb{R})$-bundle of frames on M we see that the question essentially includes within it the following **general problem.**

[1] Research partially supported by NSF Grant DMS-8301882.

Problem 2. Understand the homomorphisms $\Gamma \to \mathrm{Diff}(M)$ where M is a compact manifold.

Taking $G \subset GL(n,\mathbb{R})$ and P to be a reduction of the frame bundle to G (i.e. a "G-structure" on M) this also includes the following variation.

Problem 3. Understand the actions of Γ on a compact manifold that preserve a G-structure.

One can of course ask these questions for any group Γ, and the cases in which the group is compact, $\mathbb{Z}$, or $\mathbb{R}$ have been studied for many years. As we shall see here, the situation for lattices in semisimple groups is governed by a different set of phenomena, namely a generalization of the Mostow-Margulis rigidity phenomena to the geometric setting. Most of the results we discuss here will require the hypothesis that Γ preserve a finite volume, and some the hypothesis that Γ acts ergodically.

Before stating results concerning these problems it is appropriate to remark that it is natural to first consider the analogous questions for actions of H. In this direction we have the following result.

Theorem. Let M be a compact manifold and $P \to M$ a G-structure where G is a real algebraic group defining a volume density on M. Let H be a connected non-compact simple Lie group and suppose that H acts non-trivially on M so as to preserve the G-structure. Then there is an embedding of Lie algebras $\mathfrak{h} \to \mathfrak{g}$.

This result first appears in [28] for $\mathbb{R}$-rank$(H) \geq 2$, and then, via a much simpler proof, in [29]. In the spirit of the Mostow-Margulis theory, we then formulate the following conjecture.

Conjecture I. Let H be a connected semisimple Lie group with finite center such that $\mathbb{R}$-rank$(H') \geq 2$ for every simple factor H' of H. Let $\Gamma \subset H$ be a lattice. Let M be a compact n-manifold and $G \subset GL(n,\mathbb{R})$ a real algebraic group defining a volume density. Suppose Γ acts smoothly on M, preserving a G-structure. Then either:

i) There is a non-trivial Lie algebra homomorphism $\mathfrak{h} \to \mathfrak{g}$;

or ii) There is a smooth Γ-invariant Riemannian metric on M.

We remark that by setting $G = SL(n,\mathbb{R})$, Conjecture I implies that every volume preserving action of Γ on a sufficiently low-dimensional compact manifold is isometric.

A number of our main results are in the direction of this conjecture. In particular, we shall see:

Theorem 5.4. Assume G is of finite type (in the sense of E. Cartan [18]; see Definition 2.1 below). Then Conjecture I is true.

We recall, for example, that $G = O(p,q)$ is of finite type while $G = SL(n,\mathbb{R})$ and $Sp(2m,\mathbb{R})$ are not. This theorem is also true for higher order G-structures, thus for example, for connections. From the work of Margulis ([9], and Theorems 3.8, 3.9 below), it follows that for every H, there is a computable integer $n(H)$ such that every isometric action of Γ on a compact manifold M with $\dim M \leq n(H)$ is an action by a finite quotient of Γ. For example, $n(SL(n,\mathbb{R})) = n-1$. Thus, we obtain from Theorem 5.3 the following sample consequence.

Corollary. Suppose $\Gamma \subset SL(n,\mathbb{R})$ is a lattice, $n \geq 3$. Let M be a compact manifold, $\dim M \leq n-1$, and suppose Γ acts on M so as to preserve a volume density, and any G-structure of finite type. Then the action of Γ is an action by a finite quotient group of Γ.

If Conjecture I were true, this corollary would then be true without the assumption that Γ preserves a G-structure.

Another class of G-structures for which the conjecture is true is that of distal G-structures (which are not generally of finite type). We recall that a real algebraic group $G \subset GL(n,\mathbb{R})$ is called distal if its reductive component is compact. (See section 9 for a variety of alternative characterizations.)

Theorem 9.2. Assume the hypotheses of Conjecture I. Suppose that G is distal and that Γ acts ergodically. Then the Γ-action is isometric (and hence Conjecture I holds.)

A natural generalization of the class of G-structures of finite type are elliptic G-structures (e.g. almost complex structures).

Theorem 11.1. Suppose G is elliptic and the automorphism group of the G-structure is transitive on M. Then Conjecture I is true.

A natural generalization of the notion of a distal structure preserving action is the notion of an N-distal action where $N \subset H$ is a closed subgroup. Namely, if Γ acts on M, we say the action is N-distal if Γ preserves a distal structure on the vector bundle $p^*(TM)$, where $p: M \times H/N \to M$ is projection. Every action is clearly $\{e\}$-distal, and being H-distal is the same as preserving a distal structure on M. We conjecture (see Section 13) that every volume preserving action of Γ is N-distal where $N \subset H$ is the unipotent radical of a minimal parabolic subgroup. This can be reformulated as a conjecture about semisimple groups acting on vector bundles. (See Section 12 for a discussion.) We then have the following result.

Theorem 14.1. For each integer $n \in \mathbb{Z}^+$, there is an integer $\ell(n)$ with the following property. Suppose H is a real algebraic simple Lie group with $\mathbb{R}$-rank$(H) \geq \ell(n)$. Then for any compact manifold M with $\dim M \leq n$, and any cocompact lattice $\Gamma \subset H$, there is no volume preserving ergodic action of Γ on M which is N-distal (where N is the unipotent radical of a minimal parabolic subgroup.)

The function ℓ is easily computable from the proof. The proof of Theorem 14.1 uses explicit estimates for asymptotic behavior of matrix coefficients of unitary representations due to Howe and Harish-Chandra.

We now turn to some other considerations related to Problems 1, 2, 3. A natural way to try to obtain new actions of a group is to try to perturb a given action. The following result shows that if one starts with an isometric action, then under a sufficiently small perturbation one remains in this class. Of course, this is very far from true for diffeomorphisms or actions of free groups.

Theorem 10.1. Let H be a connected semisimple Lie group with finite center such that $\mathbb{R}$-rank$(H') \geq 2$ for every simple factor H' of H. Let $\Gamma \subset H$ be a lattice, and assume Γ acts smoothly on a compact manifold M preserving a smooth Riemannian metric. Fix a finite generating set $\Gamma_0 \subset \Gamma$. Then any action of Γ on M which

i) preserves a smooth volume density;

ii) is ergodic;

and iii) for elements of Γ_0 is a sufficiently small C^∞-perturbation of the original action;

actually leaves a C^∞ metric invariant.

Furthermore, with $n = \dim M$, for each n and $k \geq 0$, there is a positive integer $r = r(n,k)$ (independent of Γ and Γ_0) such that any Γ-action satisfying (i), (ii), and

(iii') for elements of Γ_0 is a sufficiently small C^r-perturbation of the original action;

leaves a C^k-Riemannian metric invariant. (For $k = 0$, we can take $r(n,0) = n^2 + n+1$. For $k \geq 1$, see Section 10 for a formula for $r(n,k)$.)

We remarked above that Conjecture I implies that volume preserving actions of Γ on a low dimensional manifold are isometric. The following result in this direction is a fairly direct consequence of Margulis' theorem and a lemma of Thurston [20]. (We state this here for $H = SL(n,R)$. See Section 3 for a more general statement.)

Theorem 3.13'. Let Γ be a lattice in $SL(n,\mathbb{R})$ $(n > 2)$ and suppose Γ acts smoothly on a connected (but not necessarily compact) manifold M with $\dim M \leq n-1$. If there is one point in M with a finite Γ-orbit, then the action is finite, i.e. factors through a subgroup of finite index.

Most of the results above have generalizations to lattices in p-adic groups and to S-arithmetic groups, and these are spelled out in the relevant sections below. A number of these results also have versions that are true for an arbitrary Kazhdan group (thus, for example, obtaining results for lattices in $Sp(1,n)$.) These are also discussed in the main body of the paper.

While the results of this paper exhibit very strong rigidity properties for actions of the discrete groups under consideration, they are clearly only a beginning in investigating the actions of these groups. Given the present state of knowledge, it is possible that one may hope to obtain a classification theorem for these actions. In particular, one can sharpen Problem 2 into a more specific question.

Problem 4. Can every smooth volume preserving action of Γ on a compact manifold be described in algebraic terms, or do there exist genuinely (differential) topological actions?

2. On jets and higher order frame bundles.

In this section we collect some basic notions of differential topology and supplement this with a few remarks concerning them that we will need. Standard references are [2], [7], [14], [15], [16].

We begin with the notion of the higher order frame bundles. Let $G(n)$ be the group of germs at $0 \in \mathbb{R}^n$ of local diffeomorphisms of $\mathbb{R}^n$ fixing 0. For $k \geq 1$, let $G_k(n)$ be the normal subgroup consisting of diffeomorphisms that agree with the identity up to order k, and let $GL(n,\mathbb{R})^{(k)}$ be the quotient $G(n)/G_k(n)$. Then $GL(n,\mathbb{R})^{(k)}$ is called the k-th order general linear group, and of course, $GL(n,\mathbb{R})^{(1)}$ is naturally isomorphic to $GL(n,\mathbb{R})$. The group $GL(n,\mathbb{R})^{(k)}$ has a natural linear realization. Namely, let $J^k(\mathbb{R}^n,0;\mathbb{R})$ be the vector space of k-jets at 0 of smooth $\mathbb{R}$-valued functions on $\mathbb{R}^n$. Then for $\varphi \in GL(n,\mathbb{R})^{(k)}$ and $f \in J^k(\mathbb{R}^n,0;\mathbb{R})$, $(\varphi,f) \mapsto f\circ\varphi^{-1}$ clearly defines a linear action of $GL(n,\mathbb{R})^{(k)}$ on $J^k(\mathbb{R}^n,0;\mathbb{R})$ and this representation is easily seen to be faithful. We have a natural embedding $GL(n,\mathbb{R}) \to GL(n,\mathbb{R})^{(k)}$, a natural projection $p_k: GL(n,\mathbb{R})^{(k)} \to GL(n,\mathbb{R})^{(1)}$, and hence a splitting $GL(n,\mathbb{R})^{(k)} \cong GL(n,\mathbb{R}) \ltimes N_k(n)$ where $N_k(n) = \ker(p_k)$. It is easy to see that $N_k(n)$ is connected.

There is a standard identification $J^k(\mathbb{R}^n,0;\mathbb{R}) \cong \sum_{j=0}^{k}{}^{\oplus}\, S^j(\mathbb{R}^n,\mathbb{R})$, where $S^j(\mathbb{R}^n,\mathbb{R})$ is the space of symmetric j-linear maps $(\mathbb{R}^n)^j \to \mathbb{R}$. (Namely, every element of $S^j(\mathbb{R}^n,\mathbb{R})$ can be considered as a homogeneous polynomial of degree j, and hence a function $\mathbb{R}^n \to \mathbb{R}$.) Under this identification, we obtain a linear representation π of $GL(n,\mathbb{R})^{(k)}$ on $\sum_{j=0}^{k}{}^{\oplus}\, S^j(\mathbb{R}^n,\mathbb{R})$ which can be described as follows. Let i be the standard representation of $GL(n,\mathbb{R})$ on $\mathbb{R}^n$. Then for $g \in GL(n,\mathbb{R}) \subset GL(n,\mathbb{R})^{(k)}$, $\pi(g) = \Sigma^{\oplus} S^j(i(g))$. For $g \in N_k(n)$, it follows readily from the definitions involved that for $f \in S^j(\mathbb{R}^n,\mathbb{R})$ we have

$$\pi(g)f = f + f' \quad \text{where} \quad f' \in \sum_{r>j}{}^{\oplus}\, S^r(\mathbb{R}^n,\mathbb{R}).$$

Therefore, $\pi(N_k(n))$ is a connected unipotent group. It follows immediately that, via π, $GL(n,\mathbb{R})^{(k)}$ is realized as a linear real algebraic group, with unipotent radical $N_k(n)$ and reductive Levi component $GL(n,\mathbb{R})$.

Now let $G \subset GL(n,\mathbb{R})$ be a real algebraic subgroup. We let $\mathcal{G}(G)$ be the group of germs of diffeomorphisms of $\mathbb{R}^n$ fixing 0, whose Jacobian matrix at 0 is in G. Let $\mathcal{G}_k(G) = \mathcal{G}(G) \cap \mathcal{G}_k(n)$, and $G^{(k)} = \mathcal{G}(G)/\mathcal{G}_k(G)$. We clearly have $G^{(1)} \cong G$, $G^{(k)} \subset GL(n,\mathbb{R})^{(k)}$, and $G^{(k)}$ is a real algebraic subgroup of $GL(n,\mathbb{R})^{(k)}$. Furthermore, $G^{(k)} \cong G \times N_k$ where $N_k = N_k(n)$ is a unipotent subgroup, and if G is reductive then N_k is the unipotent radical of $G^{(k)}$, and G is the reductive Levi component.

Suppose now that M is a manifold. Let $P^{(k)}(M)$ be the space of k-jets at 0 of local (smooth) diffeomorphisms $f:U \to M$ (where $0 \in U$). Then $GL(n,\mathbb{R})^{(k)}$ clearly acts on the right on $P^{(k)}(M)$ by $(f\cdot\varphi) = f \circ \varphi$ (where $f \in P^{(k)}(M)$, $\varphi \in GL(n,\mathbb{R})^{(k)}$.) In fact, $P^{(k)}(M)$ is a principal $GL(n,\mathbb{R})^{(k)}$-bundle, called the k-th order frame bundle, (the projection onto M simply being evaluation at 0, i.e. the 0-jet of f.) For $k = 1$, of course we obtain the usual frame bundle $P(M)$. Now let $G \subset GL(n,\mathbb{R})$ be a real algebraic subgroup, and $P \to M$ a G-structure on M, i.e. a reduction of the bundle $P(M)$ to G. Let $P^{(k)}$ be the k-jets of local diffeomorphisms of a neighborhood U of $0 \in \mathbb{R}^n$, $U \to M$, such that the induced map on the first order frame bundles takes the product G-structure on U into P. It is clear that $P^{(k)}$ is a prinicpal $G^{(k)}$-bundle, and in fact, $P^{(k)} \subset P^{(k)}(M)$ is a reduction of $P^{(k)}(M)$ to $G^{(k)} \subset GL(n,\mathbb{R})^{(k)}$. The bundle $P^{(k)}$ is called the k-th prolongation of the G-structure P.

(We remark that this is not in general equivalent to the standard notion of the prolongation of a G-structure [7]. While the standard notion could be applied throughout this paper, the present notion is simpler and suffices for our purposes.)

If $P \to M$ is a G-structure, we let $\mathrm{Aut}(P) = \{f \in \mathrm{Diff}(M) \mid f_*(P) = P\}$ where $f_*:P(M) \to P(M)$ is the induced map. The group Diff(M) acts naturally (on the left) on $P^{(k)}(M)$, and it is clear that Aut(P) leaves $P^{(k)}$-invariant. We may thus consider Aut(P) as a group of automorphisms of the principal bundle $P^{(k)}$.

We recall that there is a natural (i.e. Diff(M)-invariant) reduction of the bundle $P(P(M)) \xrightarrow{\pi} P(M)$ to a unipotent subgroup of $GL(n+n^2,\mathbb{R})$. Namely, choose a frame for $\mathfrak{gl}(n,\mathbb{R})$. Since each vertical tangent space of π can be identified with $\mathfrak{gl}(n,\mathbb{R})$, this gives a vertical frame at each point of $P(M)$. Each point $x \in P(M)$ is a frame of $TM_{p(x)}$ (where $p:P(M) \to M$) and hence there is a natural Diff(M)-invariant framing of $p^*(TM)$. Since there is a exact sequence $0 \to \mathrm{Vert}(P(M)) \to T(P(M)) \to p^*(TM)$, our assertion is clear. The following fact is a natural generalization.

Proposition 2.1. If $P \to M$ is a G-structure, there is a natural injective homomorphism of principal bundles over $P^{(k)}$, $P^{(k+1)} \to P(P^{(k)})$, commuting with the action of Aut(P). In particular, there is an Aut(P)-invariant U_{k+1}-structure on $P^{(k)}$, where $U_{k+1} = \ker(G^{(k+1)} \to G^{(k)})$. (We recall that U_{k+1} is a unipotent group.)

We recall that any $O(p,q)$-structure $P \to M$ (i.e., pseudo-Riemannian metric on M) defines in a canonical way a connection (the Levi-Civita connection) on the frame bundle of M, and hence an Aut(P)-invariant global framing on the manifold $P(M)$. The following notion (due to E. Cartan) generalizes this situation.

Definition 2.2. We call G of finite type if for every G-structure $P \to M$, there is a finite sequence of principal bundles $Q_{i+1} \to Q_i$, $1 \le i \le k$, where: i) $Q_1 = P$; ii) Q_{i+1} is an Aut(P)-invariant reduction of $P(Q_i) \to Q$ to a unipotent algebraic group; and iii) there is an Aut(P)-invariant global framing of Q_k.

Example 2.3. $O(p,q)$ is of finite type. $SL(n,\mathbb{R})$, $Sp(2m,\mathbb{R})$, $GL(n,\mathbb{C})$ ($\subset GL(2n,\mathbb{R})$) are not.

We remark that one can give a criterion for G to be of finite type purely in terms of the Lie algebra of G [7].

Suppose now that M is a manifold with a smooth Riemannian metric. Fix an inner product on the Lie algebra $\mathfrak{gl}(n,\mathbb{R})$ which is $\mathrm{Ad}_{GL(n,\mathbb{R})}(O(n,\mathbb{R}))$-invariant. The Riemannian metric defines a principal $O(n,\mathbb{R})$-bundle $P \subset P(M)$, and the above inner product on $\mathfrak{gl}(n,\mathbb{R})$ clearly yields an $O(n,\mathbb{R})$-invariant metric on the vertical subbundle $\mathrm{Vert}(P(M)) \to T(P(M)) \to P(M)$ defined over the subset P. Since $GL(n,\mathbb{R})$ acts freely on $P(M)$, it is clear that this extends to a $GL(n,\mathbb{R})$-invariant metric on $\mathrm{Vert}(P(M)) \to P(M)$ defined over all $P(M)$. On the other hand, the Riemannian metric defines a Levi-Civita connection on $P(M)$, and lifting the metric on M to the horizontal subspaces gives us a $GL(n,\mathbb{R})$-invariant metric on all $P(M)$. Letting $\mathrm{Met}(N)$ denote the space of smooth Riemannian metrics on a manifold N, one thus has a map $\mathrm{Met}(N) \to \mathrm{Met}(P(N))$, $\xi \to \xi^{(1)}$, with the following properties:

i) If $f:M \to M'$ is a diffeomorphism, and $f^{(1)}:P(M) \to P(M')$ is the induced map, then $(f^{(1)})^*(\xi^{(1)}) = (f^*\xi)^{(1)}$.

ii) $\xi \to \xi^{(1)}$ is continuous where $\mathrm{Met}(M)$ has the topology of C^k-uniform convergence on compact sets, and $\mathrm{Met}(P(M))$ has the topology of C^{k-1}-uniform convergence on compact sets.

(Loss of one derivative in (ii) comes from the use of the Levi-Civita connection, which depends upon the 1-jet of the metric.) Iterating this procedure and using Proposition 2.3, we arrive at the following.

Proposition 2.4. If $P \to M$ is a G-structure, there is a natural map $\mathrm{Met}(M) \to \mathrm{Met}(P^{(k)})$, $\xi \to \xi^{(k)}$, with the following properties:

i) $\xi^{(k)}$ is $G^{(k)}$-invariant.

ii) For any $\ell \geq k$, the map is continuous where $\mathrm{Met}(M)$ has the topology of C^{ℓ}-uniform convergence on compact sets, and $\mathrm{Met}(P^{(k)})$ has the topology of $C^{\ell-k}$-uniform convergence on compact sets.

iii) If $f:M \to M'$ is a diffeomorphism preserving G-structures P, P' with induced map $f^{(1)}:P \to P'$, then $(f^{(k)})^*(\xi^{(k)}) = (f^*(\xi))^{(k)}$.

We now consider a similar construction for vector bundles rather than the higher order frame bundles. Let N be a manifold and $E \to N$ a (smooth) fiber bundle. By $J^k(N;E)$ we will mean the space of k-jets of sections of E. Then $J^k(N;E)$ is a fiber bundle over N, and if E is a vector bundle, so is $J^k(N;E)$. For $E = \underline{\mathbb{R}}$, the trivial line bundle over N, we clearly have $J^k(N;\underline{\mathbb{R}})$ is a vector bundle with fiber $J^k(\mathbb{R}^n,0;\mathbb{R})$ $(n = \dim N)$, and in fact $J^k(N;\underline{\mathbb{R}})$ is simply the vector bundle associated to the principal bundle $P^{(k)}(N) \to N$ via the natural representation of $GL(n,\mathbb{R})^{(k)}$ on $J^k(\mathbb{R}^n,0;\mathbb{R})$. If E, F are vector bundles over N, we let $S^r(E,F)$ be the vector bundle whose fiber at $x \in N$ is $S^r(E_x,F_x)$, the symmetric r-linear maps $(E_x)^r \to F_x$. If $N \subset \mathbb{R}^n$ is open and E is a product bundle over N, then we have a standard identification

$$J^k(N;E) \cong \sum_{r=0}^{k} {}^{\oplus} S^r(TN,E).$$

However, this decomposition cannot be carried over in a natural way to arbitrary N. In general, there is a natural exact sequence

$$0 \to S^k(TN,E) \to J^k(N;E) \to J^{k-1}(N;E) \to 0$$

of vector bundles. (This is the jet bundle exact sequence.) There is no natural splitting of this sequence. (One exception is that the composition $J^k(N;\underline{\mathbb{R}}) \to J^0(N;\underline{\mathbb{R}}) \to 0$ is naturally split by the "constant maps".) However, given a connection ∇_E on E and a connection ∇_{T^*N} on T^*N, then there is a natural splitting of the jet bundle exact sequence and hence a natural (given the connections) identification $J^k(N;E) \cong \Sigma_{r=0}^k \oplus S^r(TN,E)$ (See [14, p.90].) If $f:E \to E'$ is a vector bundle isomorphism covering a diffeomorphism $N \to N'$ (which we still denote by f), then f induces natural maps $J^k(f):J^k(N;E) \to J^k(N';E')$ and $S^r(f):S^r(TN,E) \to S^r(TN',E')$, If f is connection preserving on E and T^*N, then under the above identification $J^k(f) = \Sigma_{r=0}^k \oplus S^r(f)$. This of course is no longer true if f fails to preserve the connections.

If V is a vector space, let $\mathrm{Inn}(V)$ be the space of positive definite inner products on V. If $E \to N$ is a vector bundle, we set $\mathrm{Inn}(N;E)$ to be the bundle whose fiber at $x \in N$ is $\mathrm{Inn}(E_x)$, and $\mathrm{Met}(N;E)$ to be the smooth sections of this bundle. We use $\mathrm{Met}(N)$ and $\mathrm{Met}(N;TN)$ interchangeably.

Consider once again the case $E = \mathbb{R}$. Any smooth metric ξ on N defines a connection on T^*N, and hence an isomorphism $J^k(N;\underline{\mathbb{R}}) \cong \Sigma_{r=0}^k \oplus S^r(TN,\mathbb{R})$. This isomorphism depends upon the k-th iteration of the covariant derivative defined by ξ [14, p.90] and hence upon the k-jet of the metric. The metric ξ also defines in a natural way a metric on each $S^r(TN,\mathbb{R})$, and hence, via the above isomorphism, a metric ξ_k on the bundle $J^k(N;\underline{\mathbb{R}})$. Summarizing:

Proposition 2.5. The map from $\mathrm{Met}(N) \to \mathrm{Met}(N;J^k(N;\underline{\mathbb{R}}))$, $\xi \mapsto \xi_k$, has the following properties:

i) If $f:N \to N'$ is a diffeomorphism, and $\xi \in \mathrm{Met}(N')$, then $(J^k(f))^*(\xi_k) = (f^*\xi)_k$.

ii) For any $\ell \geq k$, the map is continuous where $\mathrm{Met}(N)$ has the topology of C^ℓ-uniform convergence on compact sets and $\mathrm{Met}(N;J^k(N;\underline{\mathbb{R}}))$ has the topology of $C^{\ell-k}$-uniform convergence on compact sets.

Suppose now that $G \subset GL(n,\mathbb{R})$ is a real algebraic subgroup. Then G acts naturally on $\mathrm{Inn}(\mathbb{R}^n)$. Thus, given a G-structure $P \to M$, we can view $\mathrm{Inn}(M)$ as a bundle associated to the principal bundle P via the action of G on $\mathrm{Inn}(\mathbb{R}^n)$. As above, we let $J^k(M;\mathrm{Inn}\ M)$ be the k-jets of sections of $\mathrm{Inn}(M) \to M$. This is a bundle with fiber $J^k(\mathbb{R}^n,0;\mathrm{Inn}(\mathbb{R}^n))$ (the k-jets at 0 of functions $\mathbb{R}^n \to \mathrm{Inn}(\mathbb{R}^n)$.) The group $G^{(k)}$ acts naturally on $J^{k-1}(\mathbb{R}^n,0;\mathrm{Inn}(\mathbb{R}^n))$, and via this action $J^{k-1}(M;\mathrm{Inn}\ M)$ becomes an associated bundle of the principal $G^{(k)}$-bundle $P^{(k)} \to M$.

Proposition 2.6. There is a $G^{(k)}$-invariant Riemannian metric on the manifold $J^{k-1}(\mathbb{R}^n,0;\mathrm{Inn}(\mathbb{R}^n))$. Hence the total space of the bundle $J^{k-1}(M;\mathrm{Inn}\ M) \to M$ admits an $\mathrm{Aut}(P)$-invariant metric on its vertical tangent bundle.

(This of course is completely standard for $k = 1$.)

Proof. By Proposition 2.5, any smooth metric ξ on a neighborhood of $0 \in \mathbb{R}^n$ defines a metric on the jet bundle $J^k(\mathbb{R}^n;T\mathbb{R}^n)$ in a neighborhood of 0. The metric on $J^k(\mathbb{R}^n;T\mathbb{R}^n)_0$ depends only on the image of ξ in $J^k(\mathbb{R}^n,0;\mathrm{Inn}(\mathbb{R}^n))$ and this dependence is continuous. Thus, we have a continuous map $J^{k-1}(\mathbb{R}^n,0;\mathrm{Inn}(\mathbb{R}^n)) \to \mathrm{Inn}(J^{k-1}(\mathbb{R}^n;T\mathbb{R}^n)_0)$. Furthermore, this is a $G^{(k)}$-map for the natural actions of $G^{(k)}$ on these spaces. We have a natural linear representation $G^{(k)} \to GL(J^{k-1}(\mathbb{R}^n;T\mathbb{R}^n)_0)$, and this is a faithful rational representation of $G^{(k)}$. It follows that the action of $G^{(k)}$ on $\mathrm{Inn}(J^{k-1}(\mathbb{R}^n;T\mathbb{R}^n)_0)$ is proper. (We recall that an action of a locally compact group H is proper if for A, B compact, $\{h \in H \mid hA \cap B \neq \phi\}$ is compact.) It follows that the action of $G^{(k)}$ on $J^{k-1}(\mathbb{R}^n,0;\mathrm{Inn}(\mathbb{R}^n))$ is proper as well. However, any smooth proper action admits an invariant Riemannian metric [8, p.9].

We now turn to the construction of Sobolev spaces. If $E \to N$ is a fiber bundle, we let $C^k(N;E)$ be the space of C^k-sections, $0 \leq k \leq \infty$. For each $f \in C^\infty(N;E)$, we have the k-jet extension $j^k(f) \in C^\infty(N;J^k(N;E))$. Suppose now that E is a vector bundle and η is a metric on the vector bundle $J^k(N;E) \to N$. It will be important to consider the case in which η is only measurable. We set

$$C^\infty(N;E)_{p,k,\eta} = \{f \in C^\infty(N;E) \mid \|j^k(f)\|_\eta \in L^p(N)\}.$$

(We assume we are given some fixed volume density on N.) Then with the norm

$$\|f\|_{p,k,\eta} = (\int \|j^k(f)\|_\eta^p)^{1/p},$$

$C^\infty(N;E)_{p,k,\eta}$ is a normed linear space, and we shall denote its completion by $L^{p,k}_\eta(N;E)$. If η is only measurable, it is possible that $L^{p,k}_\eta(N;E) = 0$. Of course, if η is smooth, $L^{p,k}_\eta(N;E) \supset C^\infty_c(N;E)$, the smooth compactly supported functions. However, if η satisfies a suitable integrability condition with respect to a smooth metric, then we will still have containment of the compactly supported sections. More precisely:

Definition 2.7. Let V be a real vector space, and $\eta,\xi \in \mathrm{Inn}(V)$. Let $M(\eta/\xi) = \max\{\|x\|_\eta \mid \|x\|_\xi = 1\}$. If $E \to N$ is a vector bundle, and η,ξ are measurable metrics on E, then $M(\eta/\xi):N \to \mathbb{R}$ is the measurable function $M(\eta/\xi)(s) = M(\eta_s/\xi_s)$.

Lemma 2.8. Suppose η is a measurable metric on $J^k(N;E) \to N$, and ξ is smooth metric on the same bundle.

i) If $M(\eta/\xi) \in L^p_{loc}(N)$, then

$$C^\infty_c(N;E) \subset L^{p,k}_\eta(N;E).$$

ii) If $M(\xi/\eta) \in L^2_{loc}(N)$, then

$$L^{2,k}_\eta(N;E) \subset L^{1,k}_{\xi,loc}(N;E).$$

Proof. (i) is clear. To see (ii) suppose $f \in L^{2,k}_\eta(N;E)$. Then we can view f as a Cauchy sequence $f_r \in C^\infty(N;E)_{2,k,\eta}$. For any relatively compact open $B \subset N$,

$$\begin{aligned} &\int_B \|j^k(f_r)(x) - j^k(f_s)(x)\|_\xi dx \\ &\le \int_B M(\xi/\eta)(x)\|j^k(f_r)(x) - j^k(f_s)(x)\|_\eta dx \\ &\le \left(\int_B M(\xi/\eta)^2\right)^{1/2}\left(\int_B \|j^k(f_r)(x) - j^k(f_s)(x)\|^2 dx\right)^{1/2} \\ &\le \left(\int_B M(\xi/\eta)^2\right)^{1/2}\|f_r - f_s\|_{2,k,\eta} . \end{aligned}$$

Thus, $\{f_r|B\}$ is Cauchy in $C^\infty(B;E)_{1,k,\xi}$, and this implies (ii).

Lemma 2.8 of course indicates the importance of local integrability conditions on $M(\eta/\xi)$ for enabling one to use Sobolev space techniques.

We conclude this section with the following result of Thurston [20].

Lemma 2.9. Suppose $\Gamma \subset G(n)$ is a finitely generated (non-trivial) subgroup, and for each $\gamma \in \Gamma$, $d\gamma_0 = I$. Then there is a non-trivial homomorphism $\Gamma \to \mathbb{R}$.

For a simple proof, see [17].

3. Review of the Mostow-Margulis theory and Kazhdan's property; applications to fixed point sets and actions with a finite orbit.

We now recall two basic properties of the groups under consideration, namely, superrigidity and Kazhdan's property. We shall in fact expand the class of groups under consideration by subsuming the "S-arithmetic" framework. Our general reference for the results in this section is [30], which contains complete proofs of all major results we state but do not prove here.

Hypothesis 3.1. Let $V = \{\infty\} \cup \{\text{primes in } \mathbb{Z}\}$. As usual for $p \in \mathbb{Z}$ prime, we let $\mathbb{Q}_p$ denote the p-adic numbers, and we set $\mathbb{Q}_\infty = \mathbb{R}$. Let $S \subset V$ be a finite set. For each $p \in S$, let H_p be a connected semisimple $\mathbb{Q}_p$-group such that every $\mathbb{Q}_p$-simple factor of H_p has $\mathbb{Q}_p$-rank at least 2. Let $H = \Pi_{p \in S}(H_p)_{\mathbb{Q}_p}$, and let $\Gamma \subset H$ be a lattice. (i.e., Γ is discrete and H/Γ has a finite H-invariant measure.)

(The basic examples of such groups are of course S-arithmetic groups. Margulis' S-arithmeticity theorem asserts that every irreducible lattice in H, where H is as in 3.1, is S-arithmetic [9], [30].)

We now recall the following fundamental superrigidity of Margulis, generalizing Mostow's rigidity theorem.

Theorem 3.2 (Margulis). Let Γ be as in 3.1 and suppose each H_p is (algebraically) simply connected. Let $k = \mathbb{R}, \mathbb{C}$, or $\mathbb{Q}_p$, and let G be a connected k-simple k-group. Suppose $\pi:\Gamma \to G_k$ is a homomorphism with $\pi(\Gamma)$ Zariski dense. Then either:

a) $\overline{\pi(\Gamma)}$ is compact (where the closure is in the Hausdorff topology);

b) $k = \mathbb{R}$ or $\mathbb{C}$, and there is a smooth surjective homomorphism $\varphi:H_\infty \to G$ such that the following diagram commutes:

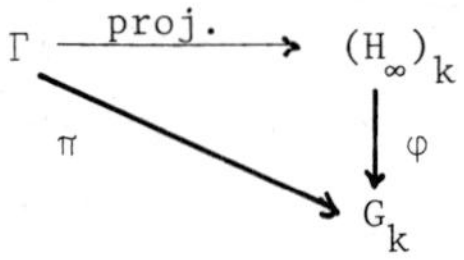

c) $k = \mathbb{Q}_p$ for some p in $S - \{\infty\}$, and there exists a k-rational surjection $\varphi: H_p \to G$ such that the corresponding diagram as in (b) commutes.

Remarks. 1) The hypothesis that each H_p be algebraically simply connected is of course not fundamental. If we drop this assumption, then the theorem is true if the conclusions in (b) and (c) are weakened to the commutativity of the diagram on $\Gamma_0 \subset \Gamma$, where Γ_0 is a subgroup of finite index. This formulation follows by lifting some subgroup $\Gamma_0 \subset \Gamma$ of finite index to a lattice in the product of the universal coverings.

2) If an irreducibility condition is assumed on Γ, it is enough to assume: $\sum_{p \in S} \mathbb{Q}_p - \operatorname{rank} H_p \geq 2$. In our present context we have chosen the above formulation so that Γ will also be a Kazhdan group.

We recall that a locally compact group Γ is a Kazhdan group if the identity is isolated in the unitary dual $\hat{\Gamma}$.

Theorem 3.3 (Kazhdan [6]) (See also [30]). If Γ is as in 3.1, then Γ is a Kazhdan group.

Corollary 3.4. For any homomorphism $\pi: \Gamma \to A$, where A is an amenable locally compact group, $\overline{\pi(\Gamma)}$ is compact.

We now collect some consequences of 3.2, 3.4 concerning representations of Γ.

Definition 3.5. Let H be as in 3.1. If $\infty \in S$, let $d(H) = \min\{\dim \pi \mid \pi$ is a non-trivial real representation of the Lie algebra $L((H_\infty)_{\mathbb{R}})\}$. If $\infty \notin S$, let $d(H) = \infty$.

Corollary 3.6. Let Γ be as in 3.1. Let $\pi: \Gamma \to GL(n,\mathbb{R})$ be a representation where $n < d(H)$. Then $\overline{\pi(\Gamma)}$ is compact.

Proof. Let L be the (real) algebraic hull of $\pi(\Gamma)$. By passing to a subgroup of Γ of finite index, one can assume L is connected. Let $L = R \ltimes U$ where R is reductive and U is unipotent. Then $R/Z(R) = \prod L_i$ where each L_i is a simple adjoint Lie group. Suppose some L_i, say L_1, is not compact. Let $q: L \to L_1$ be the projection, so that $q(\pi(\Gamma))$ is Zariski dense in L_1. By Theorem 3.2, $q \circ \pi$ extends to a rational homomorphism $(H_\infty)_{\mathbb{R}} \to L_1$. It follows that there is a non-trivial Lie algebra homomorphism $L((H_\infty)_{\mathbb{R}}) \to L(R/Z(R)) = L([R,R])$. Since $n < d(H)$, this is impossible. Thus, each L_i is compact, and it follows that R, and hence L, is amenable. An application of Corollary 3.4 completes the proof.

From Theorem 3.2 and the techniques of the proof, one can also deduce a great deal about which compact Lie groups admit a dense range homomorphism from Γ. Namely, we have the following result of Margulis. As no proof appears in the literature, we present a proof here, using [30] as the basic reference.

Definition 3.7. Suppose H is as in 3.1. Let

$$n(H) = \min\{\dim_{\mathbb{C}} H' \mid H' \text{ is a simple factor of } H\} .$$

(We recall that this is equal to

$$\varepsilon \min\{\dim_{\mathbb{R}}(L_{\mathbb{R}}) \mid L \text{ is an } \mathbb{R}\text{-simple factor of } (H_\infty)_{\mathbb{R}}\}$$

where $\varepsilon = 1/2$ or 1.)

Theorem 3.8 (Margulis). Let Γ be as in 3.1 and assume $S = \{\infty\}$. (Thus Γ is a lattice in a semisimple Lie group.) Suppose K is a compact simple Lie group and $\pi: \Gamma \to K$ is a homomorphism with $\overline{\pi(\Gamma)} = K$. Then $\dim K \geq n(H)$.

Proof. We rely heavily on the presentation of Margulis' arguments in [30, Chapter 6]. We can assume that K is an adjoint group. Write $K = L_{\mathbb{R}}$ where L is a simple $\mathbb{Q}$-group. Passing to a subgroup of Γ of finite index, we can assume L is connected. Arguing exactly as in [30, 6.1.6, 6.1.7], we see that $\mathrm{Tr}(\pi(\Gamma)) \subset \overline{\mathbb{Q}} \cap \mathbb{R}$, and that as a $\overline{\mathbb{Q}} \cap \mathbb{R}$-group, we can assume $\pi(\Gamma) \subset L_{\overline{\mathbb{Q}} \cap \mathbb{R}}$. Since Γ is finitely generated and $\pi(\Gamma)$ is Zariski dense, there is a number field k, $\mathbb{Q} \subset k \subset \bar{\mathbb{Q}} \cap \mathbb{R}$, $[k:\mathbb{Q}] < \infty$, such that $\pi(\Gamma) \subset L_k$. By restriction of scalars [30, 6.1.3] (and possibly passing to a subgroup of finite index) we see, as in [30, p. 120], that there is a semisimple $\mathbb{Q}$-group M, a k-surjection $q:M \to L$, and a homomorphism $\alpha:\Gamma \to M_{\mathbb{Q}}$ such that:

i) $\alpha(\Gamma)$ is Zariski dense in M;

and ii) $q \circ \alpha = \pi$.

Furthermore, for every $\mathbb{C}$-simple factor M_1 of M, we have $\dim M_1 = \dim L$. By Theorem 3.2, for each prime $p \in \mathbb{Z}$, the composition $\Gamma \to M_{\mathbb{Q}} \to M_{\mathbb{Q}_p}$ has bounded image. Thus, as in [30, p. 121], we deduce that a subgroup of Γ of finite index is contained in $M_{\mathbb{Z}}$. In particular, $\alpha(\Gamma)$ is discrete. Since K is not finite, $\alpha(\Gamma)$ is infinite, so $M_{\mathbb{R}}$ must be non-compact. By Theorem 3.2, the projection of α onto some $\mathbb{R}$-simple factor M_1 of M extends to an $\mathbb{R}$-rational surjection $H \to M_1$, so that M_1 is an $\mathbb{R}$-simple factor of H. If M_1 is $\mathbb{C}$-simple, $\dim_{\mathbb{C}} M_1 = \dim_{\mathbb{C}} L = \dim_{\mathbb{R}} K$, so $\dim K \geq n(H)$. On the other hand, if M_1 is not $\mathbb{C}$-simple, then $\dim_{\mathbb{C}} M_1 \geq 2n(H)$, and $\dim_{\mathbb{C}} M_1 = 2 \dim_{\mathbb{C}} L = 2 \dim_{\mathbb{R}} K$. Thus, $\dim K \geq n(H)$ in this case as well.

A similar argument, combined with that of [30, 6.1.10] shows:

Theorem 3.9 (Margulis). Let Γ be as in 3.1, $S = \{\infty\}$, and assume Γ is irreducible (i.e., projects densely in the Hausdorff topology into the connected components of the $\mathbb{R}$-points of each $\mathbb{R}$-simple factor of H.) Assume further that $H_{\mathbb{R}}/\Gamma$ is not compact. Then for any homomorphism $\pi:\Gamma \to K$, where K is a compact Lie group, we have $\pi(\Gamma)$ is finite.

All of the groups under consideration above are Kazhdan, but results of a similar type, under much more restrictive hypotheses, are valid for arbitrary Kazhdan groups.

Theorem 3.10 [27]. Let Γ be a discrete Kazhdan group and G a simple non-compact Lie group. Suppose G is not Kazhdan. Let $\pi:\Gamma \to G$ be a homomorphism. Then $\overline{\pi(\Gamma)}$ is compact.

Theorem 3.11 [27]. Let Γ be a discrete Kazhdan group and K a compact group locally isomorphic to $SO(3,\mathbb{R})$ or $SO(4,\mathbb{R})$. Let $\pi:\Gamma \to K$ be a homomorphism. Then $\pi(\Gamma)$ is finite.

We now remark that the results of this section, when combined with Lemma 2.9, yield information on fixed point sets.

Lemma 3.12. Let Γ be a Kazhdan group acting on a manifold M, with frame bundle $P(M) \to M$. Then the set of Γ-fixed points in $P(M)$ is both open and closed.

Proof. Suppose $u \in P(M)$ is a fixed point. Since Γ preserves a unipotent structure on $P(M)$, the representation of Γ on $T(P(M))_u$ is trivial by Corollary 3.4. Using 3.4 again in conjunction with Lemma 2.9, we deduce that every element of Γ is trivial in a neighborhood of u, and since Γ is finitely generated, Γ fixes a neighborhood of u.

Theorem 3.13. Let Γ be as in 3.1, $S = \{\infty\}$, $d(H)$ as in 3.5, $n(H)$ as in 3.7. Suppose Γ acts smoothly on a connected manifold M. Let $N \subset M$ be a submanifold consisting of Γ fixed points. Let $d = \mathrm{codim}_M(N)$. If

i) $d < d(H)$,

and ii) $\frac{d(d+1)}{2} < n(H)$,

then the action of Γ is finite (i.e. is trivial on a subgroup of finite index.) If we further assume Γ is as in 3.9, we need not assume (ii).

Proof. Choose $x \in N$ and let $V = TM_x/TN_x$. By 3.6, 3.8, 3.9, the representation of Γ on V is finite. Let Γ_0 be the kernel of this representation. Then Γ_0 is Kazhdan and the representation of Γ_0 on TM_x, being trivial on TN_x and V is unipotent. By 3.4, this representation is trivial. By Lemma 3.12 the action of Γ_0 is trivial.

If $\dim N = 0$, we deduce that on a connected manifold M of dimension d, with d as in (i) and (ii), an action of Γ with a fixed point must in fact be a finite action. Similar results can be obtained for arbitrary Kazhdan groups using 3.10, 3.11. Thus, we have:

Theorem 3.14. Let Γ be a Kazhdan group acting smoothly on a connected manifold M.

i) Let $N \subset M$ be a submanifold consisting of fixed points. If codim $N \leq 2$, then the Γ-action is finite.

ii) If M is a complex manifold, Γ acts holomorphically, $N \subset M$ is a complex submanifold consisting of fixed points, and $\text{codim}_{\mathbb{C}} N \leq 2$, then the Γ-action is finite.

iii) Suppose Γ acts smoothly on a surface. If there is a finite orbit, then the action is finite.

4. Superrigidity for cocycles and measurable invariant metrics.

The geometric results at the end of the preceding section were basically straightforward applications of results on finite dimensional representations. In this section we discuss a result basic for all our further considerations, namely the superrigidity theorem for cocycles. This result, first proved in [23], is a generalization of Margulis' superrigidity theorem, and in particular shows that conjecture I (of Section 1) is true if one only demands that Γ preserve a measurable invariant metric rather than a smooth one. A good part of the subsequent sections of this paper will be aimed at elucidating the relation between the existence of a measurable invariant metric and the existence of a smooth one, for actions of the groups under consideration. For complete proofs and discussion of the results in this section, see [30].

Suppose $P \to M$ is a principal G-bundle and that Γ is a group acting by principal bundle automorphisms. Let $s:M \to P$ be a measurable section (these clearly always exist) and $\Phi:M \times G \to P$ be the corresponding measurable trivialization, i.e. $\Phi(m,g) = s(m)g$. The corresponding action of Γ on $M \times G$ is given by $\gamma(m,g) = (\gamma m, \alpha(\gamma,m)g)$ where $\alpha:\Gamma \times M \to G$ is measurable and satisfies the cocycle equation: $\alpha(\gamma_1\gamma_2,m) = \alpha(\gamma_1,\gamma_2 m)\alpha(\gamma_2,m)$. Choosing a different section is equivalent to choosing an equivalent cocycle. We recall that if $\alpha,\beta:\Gamma \times M \to G$ are (measurable) cocycles, they are called equivalent if there is a (measurable) map $\varphi:M \to G$ such that $\beta(\gamma,m) = \varphi(\gamma m)^{-1}\alpha(\gamma,m)\varphi(m)$. (In the measure theoretic context it is important to allow this equation to hold only a.e. in M for each $\gamma \in \Gamma$. However, we shall for the most part disregard this point throughout the paper.)

If $E \to M$ is a vector bundle and Γ acts on E by vector bundle automorphisms, by a measurable Γ-invariant metric ξ on E we of course mean a measurable assignment, $m \to \xi_m$, of an inner product on E_m to m, which is Γ-invariant. (Once again, this only means that for each $\gamma \in \Gamma$, $(\gamma^*\xi)_m = \xi_m$ a.e., but this point too shall for the most part be ignored below.) The following is clear.

Proposition 4.1. Suppose Γ acts by principal G-bundle automorphisms of $P \to M$ and that $\alpha\colon \Gamma \times M \to G$ is an associated cocycle. If $\alpha \sim \beta$ where $\beta(\Gamma \times M) \subset K \subset G$, for some compact subgroup K, then there is a measurable Γ-invariant metric on any vector bundle $E \to M$ associated to a (finite-dimensional) linear representation of G. If G is a real algebraic group and we restrict attention to rational representations, then the converse is true.

Proposition 4.1 thus raises the question of understanding the measurable cocycles $\Gamma \times M \to G$, and the superrigidity theorem for cocycles sheds a great deal of light on them. To clarify our hypotheses, we first state the following:

Proposition 4.2. Suppose G is an algebraic group defined over k, where k is a local field of characteristic 0. Let Γ be a group acting ergodically on a measure space (M,μ) and $\alpha\colon \Gamma \times M \to G_k$ a cocycle. Then there is an algebraic k-group $L \subset G$ such that:

i) $\alpha \sim \beta$ where $\beta(\Gamma \times M) \subset L_k$;

ii) L_k is minimal with this property (among k-points of k-groups);

iii) L_k is unique up to conjugacy in G_k for groups satisfying (i), (ii).

We say that L_k (or more precisely, its conjugacy class) is the algebraic hull of α. If $L_k = G_k$, we say that α is Zariski dense in G_k.

For a proof, see [30, 9.2].

We introduce one more piece of notation. If $\pi\colon \Gamma \to G$ is a homomorphism, we let $\alpha_\pi\colon \Gamma \times M \to G$ be the cocycle $\alpha_\pi(\gamma,m) = \pi(\gamma)$.

Theorem 4.3. (Superrigidity for cocycles (cf. Theorem 3.2) [30, Theorems 5.2.5, 9.4.14, 10.1.6]. Let Γ be as in 3.1, and suppose each H_p is algebraically simply connected. Suppose (X,μ) is an ergodic Γ-space where μ is a Γ-invariant probability measure. Let $k = \mathbb{R}$, $\mathbb{C}$, or $\mathbb{Q}_a$ for some prime $a \in \mathbb{Z}$, and G a connected k-simple k-group. Suppose $\alpha : \Gamma \times X \to G_k$ is a cocycle. Then either

i) α is equivalent to a cocycle taking values in a compact subgroup of G_k;

or ii) $k = \mathbb{R}$ or $\mathbb{C}$, and there is a smooth surjective homomorphism $\varphi : H_\infty \to G$, and a cocycle $\beta \sim \alpha$ such that the following diagram commutes:

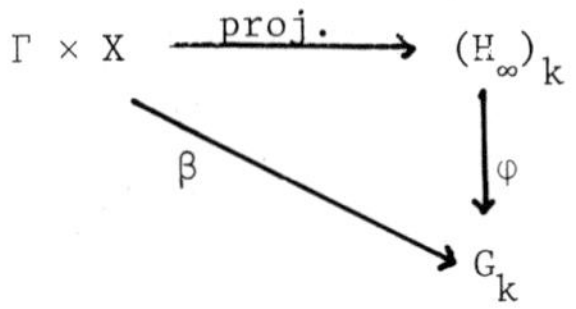

In other words $\alpha \sim \alpha_{\text{proj}\circ\varphi}$;

or iii) For some $p \in S - \{\infty\}$, $k = \mathbb{Q}_p$ and there is a k-rational surjection $H_p \to G$ such that the corresponding diagram as in (ii) commutes.

For Kazhdan groups, we have the following generalization of Corollary 3.4.

Theorem 4.4 [30]. Let Γ be a Kazhdan group and suppose Γ acts ergodically on X preserving a probability measure μ. Let $\alpha : \Gamma \times X \to A$ be a cocycle where A is amenable. Then $\alpha \sim \beta$ where $\beta(\Gamma \times X) \subset K$ for some compact subgroup $K \subset A$.

From 4.1, 4.3, 4.4, we obtain:

Theorem 4.5. Let Γ be as in 3.1 and suppose Γ acts by principal bundle automorphisms of a G-structure $P \to M$ where G is a real algebraic group. Suppose further that Γ preserves a finite volume density on M. Suppose every Lie algebra homomorphism $L((H_\infty)_{\mathbb{R}}) \to L(G)$ is trivial. Then every vector bundle $E \to M$ associated to a linear representation of G has a measurable Γ-invariant metric.

(Thus, Theorem 4.5 reduces a proof of Conjecture I to understanding the relation of measurable Γ-invariant metrics to smooth Γ-invariant metrics.)

<u>Proof</u>. By passing to the ergodic components of the Γ action on M and using 4.1, it suffices to show that the associated cocycle $\alpha:\Gamma \times M \to G$ is equivalent to a cocycle into a compact subgroup of G. We may also assume each H_p is simply connected (cf. remark 1 following Theorem 3.2). The argument is then similar in spirit to that of Corollary 3.6. Replacing G by the algebraic hull of α (Proposition 4.2), we may assume α is Zariski dense. Write $G = R \ltimes U$ where R is reductive and U is unipotent. We have $R/Z(R) = \Pi L_i$ where L_i are simple adjoint groups. The projection of α onto each L_i is a cocycle with algebraic hull L_i. Since L_i is locally isomorphic to a subgroup of $[R,R] \subset G$, the triviality of all Lie algebra homomorphisms $L((H_\infty)_{\mathbf{R}}) \to L(G)$ implies by Theorem 4.3 that each L_i is compact. Therefore G is amenable, and the theorem follows from Theorem 4.4.

<u>Corollary 4.6</u>. Let Γ be as in 3.1, and suppose Γ acts by vector bundle automorphisms on $E \to M$, where the action on M preserves a finite volume. If $\operatorname{rank}(E) < d(H)$ (Def. 3.5), then there is a measurable Γ-invariant metric on E.

<u>Remark</u>: The proof shows that 4.5, 4.6 are valid under the weaker hypothesis that (almost) every ergodic component of the action of Γ has a finite invariant measure. This allows us to apply these results to certain infinite volume manifolds (which we will need to do.)

The following observation is useful.

<u>Proposition 4.7</u>. Let Γ be a Kazhdan group and suppose Γ acts smoothly on a compact manifold M, preserving a volume density and a G-structure $P \to M$, where G is real algebraic. Let $\alpha^{(k)}$ be the cocycle corresponding to a measurable trivialization of $P^{(k)} \to M$, and α the cocycle defined by the corresponding trivialization of $P \to M$. If α is equivalent to a cocycle into a compact subgroup, so is $\alpha^{(k)}$.

<u>Proof</u>. We can write $G^{(k)} = G \ltimes N_k$ where N_k is unipotent. If we let $p:G^{(k)} \to G$ be the projection, then $p \circ \alpha^{(k)} = \alpha$. Since α is equivalent to a cocycle into $K \subset G$, $\alpha^{(k)}$ is equivalent to a cocycle into $p^{-1}(K) = K \ltimes N_k$. Since the latter group is amenable, the result follows from 4.4.

Proposition 4.7 implies in particular that for a volume preserving action of a Kazhdan group Γ on a compact manifold, if there is a measurable Γ-invariant metric on $TM \to M$, there is a measurable Γ-invariant metric on all jet bundles $J^k(M;\underline{\mathbb{R}}) \to M$.

The proof of Corollary 4.6 of course used only a part of the superrigidity theorem for cocycles. We now describe another way in which Theorem 4.3 can be used to deduce the existence of a Γ-invariant metric.

For each positive integer r and each $\gamma \in \Gamma$, let $M_r(\gamma) = \min_\pi\{\max\{|\lambda| \mid \lambda$ is an eigenvalue of $\pi(\tilde{\gamma})\}$ where π is a non-trivial representation of the (algebraic) universal covering $\tilde{H}_\infty$ on $\mathbb{C}^n$, $n \le r$, and $\tilde{\gamma}$ is an arbitrary lift of γ to $\tilde{H}_\infty$. Since there are only finitely many such objects for a given r, we have $1 \le M_r(\gamma) < \infty$. Furthermore, for each $r \ge d(H)$ (see 3.5), there is $\gamma \in \Gamma$ with $M_r(\gamma) > 1$.

<u>Theorem 4.8</u>. Let Γ be as in 3.1, and suppose Γ acts by vector bundle automorphisms of $E \to M$. Suppose further that Γ preserves a finite volume on M. Let $r = \mathrm{rank}(E)$, and ξ a smooth metric on E. If there is some $\gamma \in \Gamma$ such that $\|M(\gamma^*\xi/\xi)\|_\infty < M_r(\gamma)$, then there is a measurable Γ-invariant metric on $E \to M$.

<u>Proof</u>. We can reduce to the ergodic case by a standard argument involving the ergodic decomposition. By an argument similar to that of the proof of Theorem 4.5, either the cocycle $\alpha:\Gamma \times M \to GL(r,\mathbb{R})$ is equivalent to a cocycle into a compact subgroup (in which case we are done by Proposition 4.1), or $\alpha \sim \beta$ where for each γ and a.e. $m \in M$,

$$(1) \quad \underline{\lim}\|\beta(\gamma^n,m)\|^{1/n} \ge M_r(\gamma).$$

However, if $\delta:\Gamma \times M \to GL(r,\mathbb{R})$ corresponds to the measurable trivialization of the frame bundle of E defined by a measurable section of frames orthonormal with respect to ξ (which always exists), then for all $\gamma \in \Gamma$, $\|\delta(\gamma,m)\| = M((\gamma^*\xi)_m/\xi_m)$. Thus,

$$(2) \quad \overline{\lim}\|\delta(\gamma^n,m)\|^{1/n} \le \|M(\gamma^*\xi/\xi)\|_\infty.$$

Since $\delta \sim \beta$ one can write

$$\beta(\gamma,m) = \varphi(\gamma m)^{-1}\delta(\gamma,m)\varphi(m) \quad \text{where} \quad \varphi:M \to GL(n,\mathbb{R})$$

is measurable. Choose $A \subset M$ of positive measure on which $\|\varphi(m)\|$, $\|\varphi(m)^{-1}\|$ are bounded. For any $\gamma \in \Gamma$, and a.e. $m \in A$, we can find a sequence γ^{n_j}, $n_j \to \infty$ such that $\gamma^{n_j} m \in A$. Then

(3) $\|\beta(\gamma^{n_j},m)\| \leq C\|\delta(\gamma^{n_j},m)\|$ for some $C > 0$.

However, (1), (2), (3) together contradict our assumptions.

For Kazhdan groups, the following is an extension of the phenomenon in Theorem 4.4.

Theorem 4.9 [27]. Let Γ be a Kazhdan group acting on a measure space (M,μ), where μ is finite and invariant. Let $\alpha:\Gamma \times M \to G$ be a cocycle where G is a connected, non-Kazhdan, simple Lie group. Then α is equivalent to a cocycle into a compact subgroup.

Corollary 4.10. Suppose Γ is Kazhdan and acts smoothly on a compact manifold preserving a real algebraic G-structure where G is simple and non-Kazhdan. Then there is a measurable Γ-invariant metric on $TM \to M$.

In particular, Corollary 4.10 applies to volume preserving actions on surfaces.

We now describe one technical consequence of the existence of measurable invariant metrics that we shall need.

Lemma 4.11. Let $P \to M$ be a principal G-bundle and suppose Γ acts by principal bundle automorphisms. We also suppose the action on M is ergodic with respect to some (not necessarily smooth) quasi-invariant measure. Let $K \subset G$ be the algebraic hull of the cocycle $\alpha:\Gamma \times M \to G$ corresponding to the action. Suppose K is compact. Then there is a measurable section $\psi:M \to P$ such that the ergodic components of the action of Γ on P are exactly the sets of the form $E = \cup_{m\in M}\psi(m)Ka$ where $a \in G$. Furthermore, the corresponding measure on E (decomposing the measure on P) is $\int^{\oplus}\text{Haar}(\psi(m)Ka)\,dm$ where $\text{Haar}(\psi(m)Ka)$ is the measure defined on the submanifold $\psi(m)Ka$ as the image of Haar measure under the map $k \to \psi(m)ka$.

For the proof, see [24, Lemma 5.4] and the references there to [22]. The function ψ of course is simply a section with respect to which the associated trivialization defines a cocycle taking values in K. It follows immediately that the sets above are Γ-invariant and hence the content of the lemma is that they are ergodic.

Corollary 4.12. Suppose Γ acts by principal G-bundle automorphisms of $P \to M$ and that Γ preserves a finite measure on M. Suppose the cocycle $\alpha : \Gamma \times M \to G$ defined by the action is equivalent to a cocycle into a compact subgroup. Then (almost) all the ergodic components of the Γ-action on P have a finite invariant measure.

We shall find the following consequence of this use.

Corollary 4.13. Let $P \to M$ be a principal G-bundle and $V \to P$ a "natural" vector bundle, (i.e. a bundle on which $\mathrm{Diff}(P)$ acts naturally; e.g. tensor bundles, jet bundles, etc; see [19].) Suppose Γ acts by principal bundle automorphisms, preserving a finite volume on M, and that on (almost) every ergodic component of M, the G-valued cocycle has a compact algebraic hull. If there is a measurable Γ-invariant metric on V, then there is a measurable $\Gamma \times G$-invariant metric on V.

Proof (Sketch). For simplicity, we assume the Γ-action on M is ergodic. The general case follows by an ergodic decomposition. By a (not straightforward) averaging argument, (see [26, Lemma 2.6]), we see that if K is the algebraic hull of α, then there is a $\Gamma \times K$-invariant measurable metric on the ergodic component in Lemma 4.11 corresponding to $a = e$. Since G transitively permutes the Γ-ergodic components of P, and K is the stabilizer in G of such a component, we can transfer this measurable metric to the other ergodic components via an element of G in an essentially well-defined way. The resulting measurable metric is G-invariant by definition. Since Γ and G commute on P, this will be Γ-invariant as well.

5. Measurable invariant metrics and G-structures of finite type.

In this section we present the first "geometric" consequences of the superrigidity theorem for cocycles. We need to recall the following facts about isometries of a Riemannian manifold.

Proposition 5.1. Let N be a connected Riemannian manifold, and $I(N)$ the group of isometries. Then $I(N)$ is a Lie group which acts properly on N. Furthermore, every $I(N)$-orbit in N is closed. (See [4] for a proof.)

Theorem 5.2. Let $P \to M$ be a G-structure of finite type (Definition 2.2) where M is a compact manifold. Let Q_i be as in Definition 2.2. Suppose $H \subset \mathrm{Aut}(P)$ is a closed subgroup and that H preserves a finite measure on M. If H preserves a measurable metric on each manifold Q_i, then H is compact.

Proof: The topology on $\mathrm{Aut}(P)$ is that for which it is a Lie group acting smoothly on M. (See [7,1.3] for a discussion.) By the definition of finite type, $\mathrm{Aut}(P) \subset I(Q_k)$, and in fact it is a closed subgroup. (Here Q_k has the Riemannian metric for which the globally defined $\mathrm{Aut}(P)$-invariant framing is orthonormal.) Thus, $H \subset I(Q_k)$ is closed. By Corollary 4.12, we see by an easy induction that there is an H-invariant probability measure on Q_k. The theorem then follows from Proposition 5.1 and the following easy lemma.

Lemma 5.3. Suppose a locally compact group H acts properly on a locally compact (second countable) space X. If there is a finite H-invariant measure μ on X then H is compact.

Proof: We recall that the properness of the action means that for $A,B \subset X$ compact, $\{h \in H \mid hA \cap B \neq \phi\}$ is precompact. Let $A \subset X$ be a compact set with $\mu(A) > 0$. If H is not compact there is $h_1 \in H$ such that $h_1A \cap A = \phi$. We can then choose $h_2 \in H$ such that h_2A, h_1A, A are mutually disjoint. Continuing inductively, we can find a sequence $h_n \in H$ such that $\{h_nA\}$ are mutually disjoint. Since $\mu(A) > 0$ and $\mu(h_nA) = \mu(A)$, this clearly contradicts the finiteness of μ.

As a consequence, we can now prove Conjecture I for G-structures of finite type.

Theorem 5.4. Let Γ be as in 3.1, M a compact n-manifold with a volume density, and $P \to M$ a G-structure of finite type (where G is a real algebraic group.) Suppose Γ acts on M preserving the volume and the G-structure. Then either:

i) There is a non-trivial Lie algebra homomorphism $L((H_\infty)_{\mathbb{R}}) \to L(G)$;

or, ii) There is a smooth Γ-invariant Riemannian metric on M.

Proof: Let Γ^- be the closure of G in Aut(P) (where the topology of the latter is as in 5.2 and its proof), so that Γ^- is a Lie group. By Theorem 5.2, it suffices to see that there is a Γ^--invariant measurable metric on each manifold Q_i. The action of Γ^- on the space of measurable metrics on this manifold is clearly Borel, and hence the stabilizers in Γ^- are closed [30, 2.1.20]. Thus, it suffices to see that Γ preserves such a metric. However, this follows via an inductive argument from Theorem 4.5 (and the remark following Corollary 4.6) and condition (iii) of Definition 2.2.

Corollary 5.5. Let H be as in 3.1, and suppose M is a compact manifold of dimension $n < d(H)$ (Def. 3.5). Suppose $P \to M$ is a G-structure of finite type. If $\Gamma \subset H$ is any lattice, then any smooth action of Γ on M preserving a volume and preserving P, must preserve a smooth Riemannian metric on M. If $S = \{\infty\}$ and Γ is irreducible and not cocompact, then any such Γ-action is finite.

Proof: 5.4, 3.5, 3.9.

An examination of the proofs involved in Theorem 5.4 (and Corollary 5.5) show that they remain valid if the hypothesis that Γ preserve a volume density is weakened to the assumption that Γ preserves a finite measure.

Although we have defined a G-structure to be a sub-bundle of $P^{(1)}(M)$, one can of course define higher order G-structures, and the results of this section remain valid in this situation as well. In particular, they apply to manifolds with a connection. For example, we have:

Corollary 5.6. Let M be a compact manifold, $\dim M < n$. Then any action of $SL(n,\mathbb{Z})$ on M preserving a volume form and a connection is a finite action.

We remark that the dimension range here is sharp, as one sees from this action of $SL(n,\mathbb{Z})$ by automorphisms of $\mathbb{R}^n/\mathbb{Z}^n$.

Theorem 5.2 can also be applied to actions of an arbitrary Kazhdan group.

Theorem 5.7. Suppose G is a non-Kazhdan, almost simple, real, linear algebraic group of finite type, (e.g. $O(1,n)$.) Let Γ be a Kazhdan group acting on a compact manifold preserving a volume and a G-structure. Then there is a Γ-invariant Riemannian metric on M. If $\dim M \leq 3$, then the action is finite.

Proof. The first assertion follows from an argument similar to that of Theorem 5.4, using 4.9, 4.7. The second assertion follows from the first assertion, Theorem 3.11, and the fact that any compact Lie group K acting effectively on a compact manifold M with $\dim M \leq 3$ satisfies $\dim K \leq 6$, and hence K is locally isomorphic to a product of a torus and at most two copies of $SO(3,\mathbb{R})$.

6. A general isometry criterion .

In this section we present a basic analytic criterion for an action on a manifold to preserve a smooth Riemannian metric. This applies to arbitrary groups, not only those on which we have been focusing. The main result of this section is the following.

Theorem 6.1 [24],[26]. Let M be a compact manifold with a given volume density. Let Γ be a group acting smoothly on M, preserving volume. Suppose further that the Γ-action is ergodic. Let $G \subset GL(n,\mathbb{R})$ be a real algebraic subgroup, and suppose that Γ preserves a G-structure $P \to M$. Then:

i) If there is a Γ-invariant $f \in L^2(P^{(k)}) \cap C^0(P^{(k)})$, $f \neq 0$, then there is a Γ-invariant C^{k-3} Riemannian metric on M.

ii) If for each $k \geq 1$ such a function exists, then there is a Γ-invariant C^∞-Riemannian metric on M.

The next lemma is the first basic step in the proof.

Lemma 6.2. Let M be a second countable Hausdorff space and μ a finite measure on M which is positive on open sets. Let G be a locally compact second countable group and $Q \to M$ a (continuous) principal G-bundle. Suppose Γ acts by principal bundle automorphisms of Q covering a μ-preserving ergodic action of Γ on M. Suppose there is a Γ-invariant function $f \in L^2(Q) \cap C^0(Q)$. (We remark that left Haar measure on G defines a measure on each fiber, and hence together with μ defines a Γ-invariant measure on Q.) Then there is

i) a compact subgroup $K \subset G$;

and ii) an open Γ-invariant conull set $W \subset M$,

such that there is a continuous Γ-invariant section $\varphi: W \to Q/K$ of the natural projection of $Q/K \to M$.

Proof. Choose some measurable trivialization $Q \cong M \times G$ of Q, and let $\alpha: \Gamma \times M \to G$ be the corresponding cocycle. We then identify f as a function $f \in L^2(M \times G)$. For each $m \in M$, let $f_m \in L^2(G)$ be given by $f_m(g) = f(m,g)$. Let π be the left regular representation of G on $L^2(G)$. Then Γ-invariance of f is the assertion that $f_{\gamma m} = \pi(\alpha(\gamma,m))f_m$. The orbits of G in $L^2(G)$ approach 0 in the weak topology as $g \to \infty$, and in particular, the orbits are locally closed. Thus, the orbit space $L^2(G)/G$ is a countably separated Borel space [30, 2.1.14]. Letting $\tilde{f}_m$ be the G-orbit of f_m, we deduce that $\tilde{f}_{\gamma m} = \tilde{f}_m$ in $L^2(G)/G$. Since $L^2(G)/G$ is countably separated and the Γ-action is ergodic, it follows that $m \mapsto \tilde{f}_m$ is essentially constant [30, 2.1.11,2.2.16]. Thus, there exists $\lambda \in L^2(G)$ such that $f_m \in \pi(G)\lambda$ for almost all $m \in M$. In particular, $\lambda \in L^2(G) \cap C^0(G)$ and $\lambda \neq 0$.

Let $\mathcal{B} \to M$ be the bundle associated to $Q \to M$ via the left action of G on $L^2(G)$, so that the fiber of $\mathcal{B}$ is $L^2(G)$. Since $\pi(G)\lambda \subset L^2(G)$ is obviously G-invariant, we have an associated bundle $\mathcal{B}_\lambda \to M$ with fiber $\pi(G)\lambda$ and a natural inclusion

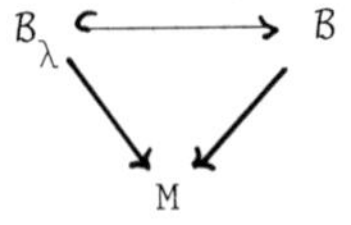

The stabilizer of λ in G, say G_λ, is compact (this is true for the regular representation of any locally compact group), and hence $\pi(G)\lambda$ is homeomorphic as a G-space to G/G_λ. We can view f as a measurable section $m \mapsto f_m$ of the bundle $\mathcal{B} \to M$ which satisfies the condition $f_m \in \mathcal{B}_\lambda$ for a.e. $m \in M$.

Now let $W = \{m \in M | f_m \in \mathcal{B}_\lambda\}$. We claim that

$W = \{m \in M | f_m \neq 0 \in L^2(G)\}$. To see this, it clearly suffices to suppose $m \notin W$, and to show $f_m = 0$. Let $U \subset M$ be an open neighborhood of m over which P is trivial. Then over U we can consider f as a function in $L^2(U \times G) \cap C^0(U \times G)$. Since W is conull, we can choose $u_n \in W \cap U$ such that $u_n \to m$. Choose $g_n \in G$ such that $f_{u_n} = \pi(g_n)\lambda$. If $\{g_n\}$ has a convergent subsequence, then on passing to a subsequence we can assume $g_n \to g$. Then $\pi(g_n)\lambda \to \pi(g)\lambda$ uniformly on compact sets, i.e. $f_{u_n} \to \pi(g)\lambda$ uniformly on compact sets. Since $u_n \to m$, we also have $f_{u_n} \to f_m$ uniformly on compact sets, so $f_m = \pi(g)\lambda$, i.e. $f_m \in \mathcal{B}_\lambda$. This contradicts the assumption that $m \notin W$. Thus we may assume $g_n \to \infty$ in G. Then $\lambda \in L^2(G)$ implies that $\pi(g_n)\lambda \to 0$ in the topology of convergence in measure on subsets of G of finite measure. However, since $\pi(g_n)\lambda = f_{u_n} \to f_m$ uniformly on compact sets, we have $f_m = 0 \in L^2(G)$, verifying our assertion.

It follows that W is an open Γ-invariant conull set. We claim that $m \to f_m$ is a continuous section of $\mathcal{B}_\lambda$ on W. To see continuity at $m \in W$, it suffices to assume $u_n \in W$, $u_n \to m$, and to show the existence of a subsequence u_{n_j} such that $f_{u_{n_j}} \to f_m$ in $L^2(G)$. As above, let $g_n \in G$ such that $f_{u_n} = \pi(g_n)\lambda$. If $g_n \to \infty$, then as above, we deduce that $f_m = 0$, contradicting the assumption $m \in W$. Let $g_{n_j} \to g \in G$. Again as above, we deduce that $f_{u_{n_j}} = \pi(g_{n_j})\lambda \to \pi(g)\lambda$ in $L^2(G)$ and that $\pi(g)\lambda = f_m$. Summarizing, $m \to f_m$ is a continuous Γ-invariant section of $\mathcal{B}_\gamma \to M$ defined on W, and to complete the proof it only suffices to recall that the fiber of $\mathcal{B}_\lambda$ is $\pi(G)\lambda$ which is homeomorphic as a G-space to G/K where $K = G_\lambda$ is compact.

We now show that in our situation W can be taken to be all of M.

Lemma 6.3. Let M be a compact manifold on which a group Γ acts ergodically by volume density preserving diffeomorphisms. Let $W \subset M$ be an open, conull, Γ-invariant set on which there is a Γ-invariant C^0-Riemannian metric. Then $W = M$. Furthermore, if we let $\overline{\Gamma}$ be the closure of Γ in the group of homeomorphisms of M preserving the topological distance function induced by the C^0-metric, then $\overline{\Gamma}$ is a compact Lie group acting continuously and transitively on M.

Proof. Choose a connected component $W_0 \subset W$. The Γ-invariant C^0-metric defines a topological distance function d on W_0 in the usual way. Γ permutes the set of connected components of W and since the measure on W is finite and invariant, ΓW_0 contains only finitely many connected components. Thus the subgroup $\Gamma_0 \subset \Gamma$ leaving W_0 invariant is of finite index. We have $\Gamma_0 \subset I(W_0,d)$, the latter being the group of isometries of W_0 with respect to d. Since W_0 is a connected manifold, $I(W_0,d)$ is locally compact [4] and the stabilizers of points in W_0 are compact. Let $\bar{\Gamma}_0$ be the closure of Γ_0 in $I(W_0,d)$. Then $\bar{\Gamma}_0$ is a locally compact isometry group with compact stabilizers, and since Γ_0 leaves the finite measure on W_0 invariant, so does $\bar{\Gamma}_0$. Since Γ is ergodic on W, Γ_0 is ergodic on W_0, and hence there is a point $x \in W_0$ with $\Gamma_0 x$ dense in W_0. It follows that $\bar{\Gamma}_0$ is transitive on W_0. (Cf. Prop. 5.1, and [4, IV. 2.2].) The finiteness of the measure on W_0 and compactness of the stabilizers imply that Haar measure on $\bar{\Gamma}_0$ is finite, and hence that $\bar{\Gamma}_0$ is compact. This implies that W_0 is compact. Ergodicity of Γ on W implies that $\Gamma W_0 = W$ so W is also compact. Since W is conull in M it is also dense, and hence $W = M$. The final assertion of the lemma follows from the fact that a locally compact group acting transitively on a (topological) manifold is a Lie group [10, p. 244].

The following two lemmas are standard.

Lemma 6.4. Let M be a compact manifold, ξ,η two C^k-Riemannian metrics on M. Let d_ξ, d_η be the corresponding distance functions. If φ is a homeomorphism of M such that $d_\xi(\varphi(x),\varphi(y)) = d_\eta(x,y)$ for all $x,y \in M$, then φ is C^{k-1}, and $\varphi^*\xi = \eta$.

Proof. This follows from the proof of [4, Theorem I.11.1].

Lemma 6.5. i) Fix $1 \leq k < \infty$. Suppose K is a compact Lie group and that K acts continuously by C^k-diffeomorphisms of a compact manifold M. Suppose further that for some smooth metric ξ, $\{g^*\xi | g \in K\}$ is precompact in the C^k-topology on C^k-metrics on M. Then there is a C^k-Riemannian metric on M invariant under K.

ii) Suppose K is a compact Lie group acting by smooth diffeomorphisms of a compact manifold M. Suppose that for some smooth metric ξ, that $\{g^*\xi | g \in K\}$ is precompact in the C^k-topology on C^k metrics for all k, $1 \leq k < \infty$. Then there is a smooth Riemannian metric on M invariant under K.

Proof. Let $\bar{\xi} = \int_{g \in K} (g^*\xi) dg$.

We can now prove Theorem 6.1.

Proof of Theorem 6.1. Fix k, $3 \leq k < \infty$. By Lemma 6.2, we can choose an open Γ-invariant conull set $W \subset M$, a compact subgroup $K \subset G^{(k)}$ and a continuous Γ-invariant section $\varphi: W \to P^{(k)}/K$. We recall that $G^{(k)} \cong G \times N_k$ where N_k is unipotent, and hence that any compact subgroup of $G^{(k)}$ is conjugate to one in G. We may thus assume that $K \subset G \subset G^{(k)}$. Let $q: P^{(k)} \to P$ be the natural projection, so that q induces a smooth map $\bar{q}: P^{(k)}/K \to P/K$. Thus, $\bar{q} \circ \varphi: W \to P/K$ is a continuous Γ-invariant section, and since K is compact, it defines a Γ-invariant C^0-Riemannian metric on W. By Lemma 6.3, $W = M$. It follows that if we let $\bar{\Gamma}$ be the closure of Γ in the group of homeomorphisms of M with the compact open topology, then $\bar{\Gamma}$ is a compact Lie group acting transitively on M. We claim that every element of $\bar{\Gamma}$ is C^{k-3} and that $\{g^*\xi | g \in \bar{\Gamma}\}$ is precompact in the C^{k-3} topology on C^{k-3} metrics, where ξ is any fixed smooth metric. By Lemma 6.5, this implies assertion (i) of Theorem 6.1. However, since k is arbitrary, Lemma 6.5 implies that assertion (ii) follows as well.

We now recall the context of Proposition 2.6. Since $K \subset G \subset G^{(k)}$, K compact, we clearly have that K fixes a point in $J^{k-1}(\mathbb{R}^n, 0; \mathrm{Inn}(\mathbb{R}^n))$, the latter being a $G^{(k)}$-space in a natural way. We thus have a $G^{(k)}$-map of $G^{(k)}/K$ onto a $G^{(k)}$-orbit in $J^{k-1}(\mathbb{R}^n, 0; \mathrm{Inn}(\mathbb{R}^n))$. It follows that we have a smooth map of bundles over M associated to $P^{(k)}$, $P^{(k)}/K \to J^{k-1}(M; \mathrm{Inn}(TM))$. Composing φ with this map, we deduce that there is a continuous Γ-invariant section $\psi: M \to J^{k-1}(M; \mathrm{Inn}(TM))$ of the natural projection. By Proposition 2.6, there is a Γ-invariant metric on the vertical tangent bundle of the fiber bundle $J^{k-1}(M; \mathrm{Inn}(TM)) \to M$. It follows that for each $m \in M$, there is a topological distance function d_m on the fiber of this bundle over m, such that d_m varies continuously in $m \in M$, and the map $m \to d_m$ is Γ-invariant. (I.e., for x,y in the fiber over m, and $\gamma \in \Gamma$, we have $d_{\gamma m}(\gamma x, \gamma y) = d_m(x,y)$.) If

$\lambda,\theta \in C^0(M;J^{k-1}(M;\text{Inn TM}))$, we define $d(\lambda,\theta) = \max\{d_m(\lambda(m),\theta(m)) \mid m\in M\}$. Then d is clearly a Γ-invariant distance function on this space of sections. The existence of the Γ-invariant section ψ implies that for any $\lambda \in C^0(M;J^{k-1}(M;\text{Inn TM}))$ and any $\gamma \in \Gamma$, we have $d(\gamma\lambda,\lambda) \le d(\gamma\lambda,\gamma\psi) + d(\psi,\lambda) \le 2d(\lambda,\psi)$. In particular, this is true for $\lambda = j^{k-1}(\xi)$ where $\xi \in C^\infty(M;\text{Inn TM})$ (i.e. a smooth metric on M.) In other words, for any such ξ, the set $\{j^{k-1}(\gamma\cdot\xi) \mid \gamma \in \Gamma\}$ is bounded in the metric d. The manifold $\text{Inn}(\mathbb{R}^n)$ is smoothly diffeomorphic to a Euclidean space $\mathbb{R}^\ell$. This enables us to define a C^∞-diffeomorphism f of Inn (TM) with a vector bundle $E \to M$ such that f is a bundle map covering the identity map on M. This induces a map $J^{k-1}(f):J^{k-1}(M;\text{Met TM}) \to J^{k-1}(M;E)$ with the property that for any smooth section ξ of Met TM, we have $J^{k-1}(f)\circ j^{k-1}(\xi) = j^{k-1}(f\circ\xi)$. (See [15, Chapter 15]. It follows that $\{j^{k-1}(f\circ(\gamma\cdot\xi)) \mid \gamma \in \Gamma\} = J^{k-1}(f)(\{j^{k-1}(\gamma\cdot\xi) \mid \gamma \in \Gamma\})$, is uniformly bounded in the space of sections $C^0(M;J^{k-1}(M;E))$. In other words, $\{f\circ(\gamma\cdot\xi) \mid \gamma \in \Gamma\}$ is bounded in the uniform C^{k-1} topology on $C^\infty(M;E)$. By the standard embedding theorems this implies that $\{f\circ(\gamma\cdot\xi) \mid \gamma \in \Gamma\}$ is precompact in $C^{k-2}(M;E)$ with the C^{k-2}-topology. Therefore $\{\gamma\cdot\xi \mid \gamma \in \Gamma\}$ is precompact in $C^{k-2}(M;\text{Inn TM})$ with the C^{k-2} topology.

Suppose now that $g \in \bar{\Gamma}$. Choose $\gamma_n \in \Gamma$ such that $\gamma_n \to g$ uniformly. By the conclusion of the preceding paragraph, by passing to a subsequence we can suppose $\gamma_n\cdot\xi \to w$ in the C^{k-2}-topology where $w \in C^{k-2}(M;\text{Inn TM})$. It follows that $g\cdot d_\xi = d_w$, where d_ξ, d_w are the corresponding distance functions. By Lemma 6.4 g is C^{k-3} and $g\cdot\xi = w$. This completes the proof.

Remarks. 1) In assertion (i) of the theorem, for $k = 1$ we can deduce the existence of a C^0-invariant metric.

2) To obtain assertion (i), the proof shows that the Γ action need not be C^∞ but only C^r for r sufficiently large.

7. The isometry criterion and integrable metrics.

In this section we show how one can obtain the hypotheses of Theorem 6.1 (and hence the conclusions) if one knows the existence of measurable invariant metrics on enough jet bundles, provided the measurable metrics have good local integrability properties.

Theorem 7.1 [24],[26]. Suppose a group Γ acts on a compact n-manifold M, preserving a G-structure $P \to M$ and a volume form. Suppose that for some $k \geq 1$ that there is a measurable Γ-invariant metric η on the vector bundle $J^r(P^{(k)};\underline{\mathbb{R}}) \to P^{(k)}$, where $r = \dim P^{(k)} + 1$, such that $M(\xi/\eta), M(\eta/\xi) \in L^2_{loc}(P^{(k)})$ for any one (and hence all) smooth metrics ξ on this bundle. (Cf. Def. 2.7.) Then there is a non-0 Γ-invariant $f \in L^2(P^{(k)}) \cap C^0(P^{(k)})$.

For the proof, we need the following lemma.

Lemma 7.2. Suppose Γ acts by principal bundle automorphisms of a principal L-bundle $Q \to N$ where L is locally compact and that the Γ action on N preserves a finite measure. Suppose there is a Γ-invariant measurable section of the map $Q/K \to N$ where $K \subset L$ is a compact subgroup. Then there is a non-zero Γ-invariant $f \in L^2(Q)$.

Proof. Let $\alpha: \Gamma \times N \to L$ be a cocycle corresponding to a measurable trivialization of L with $\alpha(\Gamma \times N) \subset K$. (This exists by the existence of a Γ-invariant section of $Q/K \to N$.) It therefore suffices to see that there is a non-zero Γ-invariant $f \in L^2(N \times L)$ under the action $\gamma \cdot (n,g) = (\gamma n, \alpha(\gamma,n)g)$. However, if $h \in L^2(L)$ is K bi-invariant, clearly $f(n,g) = h(g)$ is such a function.

Proof of Theorem 7.1. We may clearly assume that η agrees with the standard metric on the naturally split trivial line bundle $J^0(P^{(k)};\underline{\mathbb{R}}) \subset J^r(P^{(k)};\underline{\mathbb{R}})$. By Lemma 2.8, the Sobolev type space $L^{2,r}_\eta(P^{(k)};\underline{\mathbb{R}})$ contains the compactly supported functions. We have a continuous linear injection $i: L^{2,r}_\eta(P^{(k)};\underline{\mathbb{R}}) \to L^2(P^{(k)},\mathbb{R})$ defined by extending the identity map on $C^\infty_c(P^{(k)})$. Since η is Γ-invariant, the Γ action on $P^{(k)}$ induces an orthogonal representation of Γ on $L^{2,r}_\eta(P^{(k)};\underline{\mathbb{R}})$. The map i is thus an intertwining operator between orthogonal representations of Γ, and hence $i^*: L^2(P^{(k)}) \to L^{2,r}_\eta(P^{(k)};\underline{\mathbb{R}})$ also intertwines these same representations. Since $i(L^{2,r}_\eta(P^{(k)};\underline{\mathbb{R}}))$ is dense in $L^2(P^{(k)};\mathbb{R})$, i^* is injective. By Lemma 7.2, we can choose $f \in L^2(P^{(k)})$, $f \neq 0$ such that f is Γ-invariant. Thus, $F = i^*(f)$ is also non-zero and Γ-invariant. Clearly $F \in L^2(P^{(k)})$ as well. By Lemma 2.8, $F \in L^{1,r}_{\xi,loc}(P^{(k)};\underline{\mathbb{R}})$. Since ξ is smooth, the standard Sobolev embedding theorems imply $F \in C^0(P^{(k)})$, completing the proof.

It will be convenient to consider an alternate version of this theorem. Namely, from Corollary 4.13 we know that under suitable hypotheses, the existence of a measurable Γ-invariant metric on $J^r(P^{(k)};\underline{\mathbb{R}}) \to P^{(k)}$ implies the existence of a measurable $\Gamma \times G^{(k)}$-invariant metric on this bundle. On the other hand, there also exist smooth $G^{(k)}$-invariant metrics by Propositions 2.4, 2.5. For a fixed k,r and for two measurable $G^{(k)}$-invariant metrics η,ξ, $M(\xi/\eta)$ (Def. 2.7) is a $G^{(k)}$-invariant function, and hence can be considered as a function on M, which we denote by $\bar{M}(\xi/\eta)$. Then we have the following consequence of Theorem 7.1.

Theorem 7.3. Suppose a group Γ acts on a compact n-manifold, preserving a G-structure $P \to M$ and a volume density on M. Fix $k \geq 1$ and set $r = \dim P^{(k)} + 1$. Suppose there is a measurable $\Gamma \times G^{(k)}$-invariant metric η on the vector bundle $J^r(P^{(k)};\underline{\mathbb{R}}) \to P^{(k)}$, and a smooth $G^{(k)}$-invariant metric ξ on the same bundle such that $\bar{M}(\xi/\eta)$, $\bar{M}(\eta/\xi) \in L^2(M)$. Then there is a non-zero, Γ-invariant $f \in L^2(P^{(k)}) \cap C^0(P^{(k)})$ (and hence Theorem 6.1 applies).

To put our situation in perspective, we shall state formally how close the results of Section 4 come to giving us the hypotheses of Theorem 7.3 under the conditions of Conjecture I.

Theorem 7.4. Let Γ be as in 3.1. Let M be a compact n-manifold, $P \to M$ a G-structure where $G \subset GL(n,\mathbb{R})$ is real algebraic, and suppose Γ acts on M so as to preserve P and a volume form. Suppose further that every Lie algebra homomorphism $L((H_\infty)_{\mathbb{R}}) \to L(G)$ is trivial. Then for each $k \geq 1$ and $r \geq 0$, there is a measurable $\Gamma \times G^{(k)}$-invariant metric on the vector bundle $J^r(P^{(k)};\underline{\mathbb{R}}) \to P^{(k)}$ (and a $G^{(k)}$-invariant smooth one as well.)

Proof. By Theorem 4.5 on each ergodic component of the Γ-action on M the algebraic hull of the cocycle defined by a measurable trivialization of $P^{(k)} \to M$ is compact. In particular, the ergodic components of the Γ action on $P^{(k)}$ have finite Γ-invariant measure (Cor. 4.12). By Corollary 4.13, it suffices to see that Γ preserves a measurable metric on $J^r(P^{(k)};\underline{\mathbb{R}}) \to P^{(k)}$. However, the Γ action on $P^{(k)}$ preserves a unipotent structure on $P^{(r)}(P^{(k)})$ (Proposition 2.3), and hence the required assertion follows from Theorem 4.4 (and an ergodic decomposition argument.)

Comparing Theorems 7.3 and 7.4 (and using Theorem 6.1) we see that the essential obstruction remaining to proving Conjecture I is the integrability hypothesis of Theorem 7.3. It is to this question that we now turn.

8. Growth estimates and integrability of measurable metrics

In this section we present some general conditions under which we can ensure that $\bar{M}(\xi/\eta)$, $\bar{M}(\eta/\xi) \in L^2(M)$ (where the notation is as in Theorem 7.3.) We shall do this by examining the growth of these functions along Γ-orbits, and then show how this can be converted into the desired integrability assertion by an application of Kazhdan's property. We remark that Kazhdan's property can be considered as a statement about the matrix coefficients for unitary representations of Γ. Finer results about these coefficients are available, and these in turn lead to some sharper results than those we discuss in this section. They will be developed in Sections 12-14.

We first make the following elementary observation. Let Λ be a group acting by vector bundle automorphisms of a vector bundle $E \to N$. Suppose η is a measurable Λ-invariant metric on this bundle, and ξ is an arbitrary smooth metric. Suppose further that for each $h \in \Lambda$, $M(h^*\xi/\xi)$ and $M(\xi/h^*\xi)$ are uniformly bounded functions on N. Then for $x \in N$ and $h \in \Lambda$, we have

$$\begin{aligned} M(\eta/\xi)(hs) &= M(h^*\eta/h^*\xi)(s) \\ &\leq M(\eta/\xi)(s)M(\xi/h^*\xi)(s). \end{aligned}$$

(Here we have of course used the invariance of η.) Thus we see that the growth of $M(\eta/\xi)(s)$ along a Λ-orbit is governed by the growth of $\|M(\xi/h^*\xi)\|_\infty$. Similarly, the growth of $M(\xi/\eta)(s)$ along a Λ-orbit is governed by the growth of $\|M(h^*\xi/\xi)\|_\infty$. The following two results are then clearly germane.

Theorem 8.1 [24]. Let Γ be a discrete Kazhdan group and $\Gamma_0 \subset \Gamma$ a finite symmetric generating set. Then there is $K > 1$ with the following property. Let (X,μ) be a measurable ergodic Γ-space where μ is finite and invariant. If $f:X \to \mathbb{R}$ is measurable and for all $\gamma \in \Gamma_0$, $|f(\gamma x)| \leq K|f(x)|$ for a.e. $x \in X$, then $f \in L^2(X)$.

Theorem 8.2. Let Γ, Γ_0, and (X,μ) be as in 8.1. Suppose p is a real polynomial. If $f:X \to \mathbb{R}$ is measurable and satisfies $|f(\gamma x)| \leq |p(n)||f(x)|$ for all $\gamma \in (\Gamma_0)^n$, all $n \in \mathbb{Z}^+$, and a.e. $x \in X$, then $f \in L^q(X)$ for all q, $1 \leq q < \infty$.

We preface the proofs of 8.1, 8.2 with a few remarks. Kazhdan's property for discrete Γ can be reformulated as follows. Let $\Gamma_0 \subset \Gamma$ be a finite generating set (which always exists for Kazhdan groups). Then there exists $\varepsilon > 0$ such that for any unitary representation π of Γ on a Hilbert space H with no (non-trivial) invariant vectors, and any unit vector $v \in H$, there is some $\gamma \in \Gamma_0$ such that $\|\pi(\gamma)v-v\| \geq \varepsilon$. Suppose now that (X,μ) is as in 8.1, 8.2 with $\mu(X)=1$. Let $H = \mathbb{C}^{\perp} \subset L^2(X)$. By ergodicity, the natural representation π on H has no (non-trivial) invariant vectors. For any $A \subset X$, let χ_A be the characteristic function, p_A the projection of χ_A onto H, and $f_A = p_A/\|p_A\|$ when A is neither null nor conull. If $A,B \subset X$ with $\mu(A) = \mu(B)$, then a straightforward calculation shows that $\|f_A - f_B\|^2 = \mu(A \Delta B)/\mu(A)(1 - \mu(A))$. So we obtain:

Lemma 8.3. Let Γ, Γ_0, (X,μ) be as in 8.1, 8.2 with $\mu(X) = 1$, and let ε be as above. Let $A \subset X$ be neither null nor conull. Then there is $\gamma \in \Gamma_0$ such that $\mu(\gamma A \Delta A) \geq \varepsilon^2\mu(A)(1 - \mu(A))$, and hence $\mu(\gamma A \cap (X - A)) \geq \frac{\varepsilon^2}{2}\mu(A)(1 - \mu(A))$.

Proof of Theorem 8.1. Choose a measurable set $A_0 \subset X$ such that $\mu(A_0) \geq 1/2$ and $|f|$ is bounded on A_0, say $|f(x)| \leq B$ for $x \in A_0$. By Lemma 8.3, we can choose $\gamma_0 \in \Gamma_0$ such that

$$\mu(\gamma_0 A_0 \cap (X - A_0)) \geq \frac{\varepsilon^2}{2}(1 - \mu(A_0)).$$

Then choose $A_1 \subset \gamma_0 A_0 \cap (X - A_0)$ such that $\mu(A_1) = (\varepsilon^2/4)(1 - \mu(A_0))$. Repeat the argument applied to $A_0 \cup A_1$. We then have $A_2 \subset \gamma_1(A_0 \cup A_1) \cap (X - A_0 \cup A_1)$ with $\mu(A_2) = (\varepsilon^2/4)(1 - \mu(A_0 \cup A_1))$. Continuing inductively we find a disjoint collection of measurable sets, $\{A_i\}$ such that $A_n \subset \Gamma_0(\cup_{j<n} A_j)$ and, setting $a_n = \mu(A_n)$, we have for $n \geq 0$ that $a_{n+1} = (\varepsilon^2/4)(1 - \Sigma_{i=0}^{n} a_i)$. Thus, if $n \geq 1$,

$$a_{n+1} = \frac{\varepsilon^2}{4}(1 - \sum_{i=0}^{n-1} a_i - a_n)$$
$$= a_n - \frac{\varepsilon^2}{4} a_n$$
$$= a_n(1 - \frac{\varepsilon^2}{4}).$$

(For $n = 0$, we have $a_1 = \varepsilon^2/4(1 - a_0)$.) We clearly have $\Sigma_{i=0}^{\infty} a_i = 1$. Suppose now that f satisfies $|f(\gamma x)| \le K|f(x)|$ for $\gamma \in \Gamma_0$ and a.e. $x \in X$. Since $A_n \subset \Gamma_0(\cup_{j<n} A_j)$, we clearly have $|f(x)| \le K^n B$ for $x \in A_n$. Then $\int |f|^2 \le \Sigma a_n K^{2n} B^2$, and this will be finite if $\overline{\lim}\ a_{n+1} K^2/a_n < 1$. Thus, any $K > 0$ with $1 < K^2 < 1/(1-\varepsilon^2/4)$ suffices.

Proof of Theorem 8.2. This follows by a similar argument as one obtains an estimate of the form $\int |f|^q \le \Sigma_{n=0}^{\infty} B^q |p(n)|^q a_n$, and since $\overline{\lim}\ a_{n+1}/a_n < 1$, this is finite.

9. Applications to distal G-structures

We recall the following definition.

Definition 9.1. Let G be a real algebraic group. Then G is called distal if the reductive Levi component of G is compact.

We also recall that the following assertions are equivalent on a real algebraic linear subgroup $G \subset GL(n,\mathbb{R})$ [11]:

i) G is distal.

ii) G acts distally on $\mathbb{R}^n$ (i.e. $g_n(v) \to 0$ for some sequence $g_n \in G$ implies $v = 0$).

iii) All eigenvalues of elements of G satisfy $|\lambda| = 1$.

iv) For any compact symmetric neighborhood V of e in G, $\max\{\|g\| \mid g \in V^n\}$ has polynomial growth in n.

v) If G^0 is the connected component of the identity, then there is a G^0-invariant flag $\{0\} \subset V_1 \subset \cdots \subset V_k = \mathbb{R}^n$ such that the induced action of G on V_i/V_{i-1} has an invariant inner product for each $i = 1,\ldots,k$.

In this section we will prove Conjecture I for distal G-structures under the additional hypothesis of ergodicity. In fact, we can obtain the result for arbitrary Kazhdan groups.

Theorem 9.2 [26]. Let Γ be a discrete Kazhdan group acting smoothly and ergodically on a compact manifold M. Suppose Γ preserves a smooth distal G-structure on M. Then Γ leaves a smooth Riemannian metric invariant.

We begin the proof by recording the following two easily verified assertions.

Lemma 9.3. If G is a distal linear group, so is the prolongation $G^{(k)}$.

Lemma 9.4. Suppose a group Γ acts on a manifold M preserving a distal G-structure $P \to M$. Then for any k, there is a $\Gamma \times G^{(k)}$-invariant distal structure on the manifold $P^{(k)}$.

The next lemma is a version of Theorem 7.4.

Lemma 9.5. Under the hypotheses of Theorem 9.2, for each $k \geq 1$ and $r \geq 0$, there is a measurable $\Gamma \times G^{(k)}$-invariant metric on the vector bundle $J^r(P^{(k)};\underline{\mathbb{R}}) \to P^{(k)}$ (and a smooth $G^{(k)}$-invariant metric as well.)

Proof. This follows via the argument in the proof of Theorem 7.4, using Theorem 4.4 in place of Theorem 4.5.

Proof of Theorem 9.2. Let $\tilde{P} \subset P(P^{(k)})$ be the $\Gamma \times G^{(k)}$-invariant distal L-structure on $P^{(k)}$ of Lemma 9.3, where L is a distal linear group. Then for each $r \geq 1$, the vector bundle $J^r(P^{(k)};\mathbb{R}) \to P^{(k)}$ is a bundle associated to $\tilde{P}^{(r)}$ via a rational representation π of $L^{(r)}$, and we thus have a $\Gamma \times G^{(k)}$-invariant distal structure on the vector bundle $J^r(P^{(k)};\underline{\mathbb{R}}) \to P^{(k)}$. The bundle $J^r(P^{(k)};\underline{\mathbb{R}}) \to P^{(k)}$ is isomorphic (but not naturally) to a direct sum of tensor bundles. Since $TP^{(k)}$ admits a globally defined frame, so does $J^r(P^{(k)};\underline{\mathbb{R}})$. We can clearly choose this frame to lie in the $\Gamma \times G^{(k)}$-invariant $\pi(L^{(r)})$-structure on $J^r(P^{(k)};\underline{\mathbb{R}})$ described above.

Let η be a measurable $\Gamma \times G^{(k)}$-invariant metric on $J^r(P^{(k)};\underline{\mathbb{R}})$ and ξ a smooth $G^{(k)}$-invariant metric (Lemma 9.5). By Theorems 7.3, 6.1, it suffices to prove that $\bar{M}(\eta/\xi)$, $\bar{M}(\xi/\eta) \in L^2(M)$, and by

Theorem 8.2 it suffices to find a real polynomial such that $\overline{M}(\eta/\xi)(\gamma m) \leq |p(n)|\overline{M}(\eta/\xi)(m)$ for all $\gamma \in \Gamma_0^n$ and all $m \in M$, (and similarly for $\overline{M}(\xi/\eta)$.) We have, for $\gamma \in \Gamma$, $g \in G^{(k)}$, and $y \in P^{(k)}$,

$$M(\eta/\xi)(\gamma y g) = M(d(\gamma,g)^*\eta/d(\gamma,g)^*\xi)(y)$$
$$\leq M(d(\gamma,g)^*\eta/\xi)(y)M(\xi/d(\gamma,g)^*\xi)(y).$$

Therefore:

(*) $M(\eta/\xi)(\gamma y g) \leq M(\eta/\xi)(y)M(\xi/d(\gamma,g)^*\xi)(y).$

Choose a subbundle $Q \subset P^{(k)}$ where Q is a reduction of $P^{(k)}$ to a maximal compact subgroup $K \subset G^{(k)}$. (This always exists since $G^{(k)}/K$ is contractible.) Of course, Q itself is compact. Each metric ξ_y, $y \in Q$ can be expressed as a matrix in terms of the basis for $J^r(P^{(k)};\underline{R})_y$ described above. Since Q is compact, $\{\xi_y | y \in Q\}$ forms a compact set of matrices. By (*) it suffices to show there is a polynomial p such that for each $\gamma \in \Gamma_0^n$ and each $y \in Q$, there is an element $g \in G^{(k)}$ such that $\gamma y g \in Q$ and $\|d(\gamma,g)_y^{-1}\| \leq |p(n)|$, where the norm is once again taken with respect to the above framing.

Let $\tilde{Q} = Q \cup \bigcup_{\gamma_i \in \Gamma_0} \gamma_i Q$, so that $\tilde{Q} \subset P^{(k)}$ is also compact. Since the action of $G^{(k)}$ on $P^{(k)}$ is proper, there is a compact subset $A \subset G^{(k)}$ such that for $g \in G^{(k)} - A$, $\tilde{Q}g \cap \tilde{Q} = \phi$. It follows that for each $y \in Q$ and each $\lambda \in \Gamma_0$, there is some $g \in A$ with $\lambda y g \in Q$. Thus, if $\lambda = \gamma_n \cdots \gamma_1 \in \Gamma_0^n$, and $y \in Q$, we can choose $g_i \in A$ and $y_i \in Q$ such that $y_1 = y$ and $\gamma_i y_i g_i = y_{i+1}$. Let $g = \Pi g_i$. Then $\gamma y g \in Q$. Identifying Γ and $G^{(k)}$ as subgroups of the direct product, one can write $\gamma g = \Pi g_i \gamma_i$, and hence $d(\gamma g)_y = \Pi_{i=1}^n (dg_i)_{z_i} (d\gamma_i)_{y_i}$ where $y_i \in Q$, $z_i \in \tilde{Q}$, $g_i \in A$. It follows that there is a compact symmetric neighborhood of the identity, $V \subset L^{(r)}$ such that (with respect to the global framing above), $d(\gamma g)_y \in V^n$ for all $y \in Q$ and $\gamma \in \Gamma_0^n$. Since $L^{(r)}$ is distal, the required polynomial exists.

Theorem 9.2 suggests the following conjecture. We recall that an action of a group Γ on a compact metric space (M,d) is called distal if $x,y \in M$, $x \neq y$, implies $\inf\{d(\gamma x, \gamma y)) | \gamma \in \Gamma\} > 0$.

Conjecture 9.6. Suppose Γ is a Kazhdan group acting smoothly on a compact manifold M. Suppose the action is distal. Then the action is isometric (with respect to an equivalent metric.)

For a discussion, see [26]. For related results, see [1], [12].

10. Applications to perturbations

In this section we show, roughly speaking, that perturbations of isometric actions are isometric. More precisely:

Theorem 10.1 [24],[26]. Let Γ be as in 3.1. Suppose Γ acts smoothly on a compact n-manifold preserving a smooth Riemannian metric. Fix a finite symmetric generating set $\Gamma_0 \subset \Gamma$. Then any action of Γ on M which:

i) preserves a smooth volume density;

ii) is ergodic;

and iii) for elements of Γ_0 is sufficiently C^∞-close to the original action;

actually leaves a smooth Riemannian metric invariant.

Furthermore, for each n and each $k \geq 0$, there is a positive integer $r = r(n,k)$ (independent of Γ), such that any Γ-action satisfying (i), (ii) and

(iii') for elements of Γ_0 is sufficiently C^r close to the original action;

leaves a C^k-Riemannian metric invariant. (The function $r(n,k)$ can be given as follows:

For $k = 0$, $r(n,0) = n^2 + n + 1$. For $k \geq 1$, $r(n,k) = n + k + 5 + \dim SL(n,\mathbb{R})^{(k+3)}$.)

Proof. Let ξ be a smooth metric on M which is invariant under the original action. Then from Proposition 2.4 we see that $\xi^{(k)}$ will be a smooth metric on $P^{(k)}(M)$ such that for actions which are sufficiently small C^k-perturbations of the original action on Γ_0, we will have $M(\gamma^*(\xi^{(k)})/\xi^{(k)})$, $M(\xi^{(k)}/\gamma^*(\xi^{(k)}))$ will be uniformly close to 1 for $\gamma \in \Gamma_0$. By Theorem 4.8, for sufficiently small C^k-perturbations, the cocycle $\Gamma \times M \to GL(n,\mathbb{R})^{(k)}$ defined by the action on

$P^{(k)}(M)$ is equivalent to a cocycle into a compact group. By Corollary 4.12, the ergodic components of the Γ-action on $P^{(k)}(M)$ have finite invariant measure. Therefore, by Proposition 2.3, Theorem 4.4, and Corollary 4.13, for each ℓ there is a measurable $\Gamma \times GL(n,\mathbb{R})^{(k)}$-invariant metric on $J^{\ell}(P^{(k)};\underline{\mathbb{R}})$. Let $(\xi^{(k)})_{\ell}$ be the smooth $GL(n,\mathbb{R})^{(k)}$-invariant metric on this bundle given by Proposition 2.5. Then for a sufficiently small $C^{\ell+k}$-perturbation, we have

$$\|\bar{M}(\gamma^*(\xi^{(k)})_{\ell}/(\xi^{(k)})_{\ell})\|_{\infty},\ \|\bar{M}((\xi^{(k)})_{\ell}/(\gamma^*(\xi^{(k)})_{\ell})\|_{\infty} < K$$

where K is any preassigned real number with $1 < K$. By Theorem 8.1 and the remarks preceding it, it follows that for sufficiently small $C^{\ell+k}$-perturbations we will have $\bar{M}(\eta/(\xi^{(k)})_{\ell})$, $\bar{M}((\xi^{(k)})_{\ell}/\eta) \in L^2(M)$. By Theorems 7.3, 6.1, for $\ell = \dim P^{(k)}(M) + 1$, this implies the existence of a Γ-invariant C^{k-3}-metric on M.

The perturbation condition in the above proof was used twice. First, by superrigidity it established (via Theorem 4.8) that the ergodic components of the Γ-action on $P^{(k)}(M)$ had finite invariant measure, and second to obtain the growth estimate needed in the hypotheses of Theorem 8.1. In fact, an examination of the proof shows that this use of superrigidity is the only point of the argument that does not apply to the case of an arbitrary Kazhdan group. We can (and will) however, give an alternate argument which shows:

Theorem 10.2. Let Γ be a Kazhdan group. Then Theorem 10.1 holds for Γ.

Proof. By the proof of Theorem 10.1 and Proposition 4.7, it suffices to show that the cocycle $\alpha:\Gamma \times M \to GL(n,\mathbb{R})$ defined by the action on $P(M)$ is equivalent to a cocycle into a compact subgroup (where the action of course is an ergodic sufficiently small perturbation of the original isometric action.)

Lemma 10.3. If there is a Γ-invariant $f \in L^2(P(M))$, $f \neq 0$, then α is equivalent to a cocycle into a compact subgroup.

Proof (cf. proof of Lemma 6.2). Suppose $f \in L^2(M \times G)$ is non-0 and Γ-invariant under the action $\gamma\cdot(m,g) = (\gamma m,\alpha(\gamma,m)g)$, where for convenience we have written $G = GL(n,\mathbb{R})$. Let π be the left regular representation of G. For $m \in M$, let $f_m(g) = f(m,g)$. Then for each $\gamma \in \Gamma$, $f_{\gamma m} = \pi(\alpha(\gamma,m))f_m$ for a.e. m. It follows that $[f_{\gamma m}] = [f_m]$ in $L^2(G)/\pi(G)$, and since $L^2(G)/G$ is a countably separated Borel space and Γ act ergodically on M, we have that $[f_m]$ is essentially constant (see [30, Chapter 2]). In other words, there is some $\lambda \in L^2(G)$ such that $f_m \in \pi(G)\lambda$ for a.e. $m \in M$. The stabilizer of λ in G is compact, and hence as a G-space we have $\pi(G)\lambda \cong G/K$ where K is compact. Therefore, viewing $m \to f_m$ as a map ψ into G/K, we deduce the existence of a measurable $\psi: M \to G/K$ such that $\psi(\gamma m) = \alpha(\gamma,m)\psi(m)$. Lifting ψ to a function $\varphi: M \to G$, we obtain $\varphi(\gamma m)^{-1}\alpha(\gamma,m)\varphi(m) \in K$, verifying the lemma.

We now return to the proof of Theorem 10.2. The original Γ-action leaves a K-structure on M invariant where $K \subset G \subset GL(n,\mathbb{R})$ is compact. Fix $f \in C_c^\infty(G)$ such that $\int |f|^2 = 1$, and f is K bi-invariant. This can be transferred to each fiber of $P(M)$ to obtain a Γ-invariant function $F \in C_c^\infty(P(M))$ whose L^2-norm on each fiber is 1. In particular, for any probability measure μ on M, we have $\|F\|_2 = 1$. (We recall that a probability measure on M defines a measure on $P(M)$, since we have Haar measure on each fiber.) For a sufficiently small C^1-perturbation of the isometric action, we will have, for $\gamma \in \Gamma_0$, $|\gamma \cdot F - F|$ uniformly small and with support contained in some fixed compact set (slightly larger than $\mathrm{supp}(F)$). It follows that for a sufficiently small perturbation, $\|\gamma \cdot F - F\|_2 < \varepsilon$ for any preassigned $\varepsilon > 0$, for all $\gamma \in \Gamma_0$. By Kazhdan's property, this implies that there is a Γ-invariant $h \in L^2(P(M))$, $h \neq 0$, and by Lemma 10.3 and the first paragraph of the proof, this suffices.

11. Applications to elliptic G-structures

We recall that a G-structure $P \to M$ is elliptic if the infinitesimal automorphisms of P are characterized as those vector fields satisfying an elliptic partial differential equation. This is equivalent to the condition that the linear Lie algebra $L(G) \subset \mathfrak{gl}(n,R)$ contains no matrices of rank 1. [7, Prop. I.1.4]. For an elliptic G-structure, $\mathrm{Aut}(P)$ is a Lie group. Finite type structures are elliptic, but the converse is false in general, a basic example being almost complex structures (i.e. $GL(n,\mathbb{C}) \subset GL(2n,\mathbb{R})$).

Theorem 11.1 [32]. Let Γ be as in 3.1. Suppose M is a compact manifold and $P \to M$ is a G-structure such that $\mathrm{Aut}(P)$ is a Lie group acting transitively on M. Suppose Γ acts on M preserving a volume density and P. Then the conclusion of the conjecture is true. I.e., if every Lie algebra homomorphism $L((H_\infty)_{\mathbb{R}}) \to L(G)$ is trivial, then there is a smooth Γ-invariant Riemannian metric on M.

Proof. By Theorem 4.5, there is a measurable Γ-invariant metric η on $TM \to M$. Let ξ be a smooth metric. By Theorem 8.1 and the paragraph preceding it, for some $\alpha > 0$, we have on each ergodic component of the Γ-action on M that $M(\eta/\xi) \in L^{2\alpha}$. Multiplying η by a measurable Γ-invariant function, we may therefore assume that with respect to the smooth Γ-invariant metric defined by the volume density that $M(\eta/\xi) \in L^{2\alpha}(M)$. Now let Y be the set of (globally defined) infinitesimal automorphisms of P. Then $\dim Y < \infty$ and for each $m \in M$, the evaluation map $e_m: Y \to TM_m$ is surjective. Define $\varphi: Y \to \mathbb{R}$ by $\varphi(F) = \int_M \|F(m)\|^{\alpha}_{\eta(m)}\, dm$. Since $M(\eta/\xi) \in L^{2\alpha}(M)$, $0 \le \varphi(F) < \infty$, and it is clear that $\varphi(F) = 0$ if and only if $F = 0$. Furthermore, φ is continuous (the topology on Y being uniquely determined since $\dim Y < \infty$) and φ is homogeneous of degree α. Therefore $\{F \in Y \mid \varphi(F) < 1\}$ is a non-empty open set with compact closure. Since η is Γ-invariant, so is φ, and hence the representation of Γ on Y is uniformly bounded. Since $\dim Y < \infty$, there is a Γ-invariant inner product on Y, and via the surjective maps $\{e_m\}$, this defines a smooth Γ-invariant metric on TM.

12. A conjecture on semisimple groups acting on vector bundles and N-distal actions

The next three sections of this paper are devoted to a proof of Conjecture I in certain dimension ranges, and for cocompact lattices, under the hypothesis of N-distality which we shall explain presently. Conjecturally, every volume preserving action (for Γ as in 3.1) satisfies this condition. We begin with a related conjecture concerning actions of the ambient semisimple group. For simplicity, we restrict ourselves to real groups, i.e. we assume the hypotheses of 3.1, but with $S = \{\infty\}$. However, everything we say can be suitably modified to apply to the general situation of 3.1.

Thus, suppose H is a connected semisimple real algebraic group such that every simple factor has $\mathbb{R}$-rank at least 2. Let M be a manifold and $E \to M$ a vector bundle over M of rank n. Suppose H acts smoothly by vector bundle automorphisms of E. Let $N \subset H$ be the unipotent radical of a minimal parabolic subgroup of H.

Definition 12.1. If L is a subgroup of H, the action of H on $E \to M$ is called L-distal if there is an L-invariant reduction of the structure group of E to a distal subgroup of $GL(n,\mathbb{R})$ (Def. 9.1.)

Conjecture 12.2. If M is compact and H preserves a volume density on M, then the action on $E \to M$ is N-distal.

This conjecture can be verified in many special cases.

Two possible weakenings of this conjecture which may be relevant to our considerations are to assume ergodicity of the H-action, or only to require the existence of such an N-invariant structure on an open N-invariant subset.

The next theorem represents some general empirical evidence for Conjecture 12.2, and is a consequence of superrigidity for cocycles.

Theorem 12.3. Assume the above hypotheses and that the H action is ergodic. Then the action is measurably N-distal. I.e., there is a measurable N-invariant reduction of the structure group to a distal group.

Proof. We may clearly assume that H is algebraically simply connected. Let $\alpha: H \times M \to GL(n,\mathbb{R})$ be the associated cocycle, and $G \subset GL(n,R)$ the algebraic hull (Proposition 4.2). Then by passing to a finite covering of M if necessary, we can assume G is Zariski connected [30, 9.2.6]. Write $G = L \ltimes U$ where L is reductive and U is unipotent. The composition of α with the projection of G onto $L/[L,L]$ is Zariski dense. Hence, by Theorem 4.4, $L/[L,L]$ is compact. Therefore $Z(L)$ is compact. Now let $L/Z(L) = L_1 \times L_2$ where L_1 is compact and L_2 is a semisimple adjoint group with no compact factors. By Theorem 4.3, by passing to an equivalent cocycle, we can assume the composition of α with the projection $p: G \to L_2$ is of the form $p \circ \alpha(h,m) = \pi(h)$ for some rational homomorphism $\pi: H \to L_2$. Since H is simply connected, we can lift π to a rational representation $\tilde{\pi}:H \to L$. It follows that $\tilde{\pi}(N)$ is unipotent. We easily deduce that $\alpha(N \times M) \subset ((K \ltimes V) \ltimes U)$ where $K \ltimes V \subset L$, V is unipotent, and K is compact. Thus, $(K \ltimes V) \ltimes U$ is distal. The proof is complete, except that we have passed to a finite cover. However, a somewhat technical argument, which we omit, shows that this suffices.

The next result is a general one about semisimple group actions.

<u>Theorem 12.4</u>. Assume the hypotheses of Conjecture 12.2. Fix a smooth metric ξ on $E \to M$. For $h \in H$, let $M(h) = \max\{M(h^*\xi/\xi)(m), M(\xi/h^*\xi)(m) \mid m \in M\}$. Then the function $f:L(N) \to \mathbb{R}$ on the Lie algebra $L(N)$ given by $f(X) = M(\exp X)$ has polynomial growth on $L(N)$.

<u>Proof</u>. The assertion is independent of ξ, so we may assume that ξ is K-invariant, where $K \subset H$ is a maximal compact subgroup. Then $M(h)$ is clearly K bi-invariant. Let $\mathfrak{h} = \mathfrak{k} \oplus \mathfrak{p}$ be the Cartan decomposition of the Lie algebra of H with respect to K. Choose a positive definite inner product on $\mathfrak{k} \oplus \mathfrak{p}$ which is $Ad_H(K)$-invariant and $\mathfrak{k} \perp \mathfrak{p}$. This defines a Riemannian metric on H/K, and we let $d(\cdot,\cdot)$ be the corresponding distance function. It is well-known (see [4], for example) that for $Y \in \mathfrak{p}$, $d(\exp(Y)[e],[e]) = \|Y\|$, and that the map $\mathfrak{p} \to H/K$, $Y \to \exp(Y)([e])$ is a diffeomorphism. The function M is continuous and clearly for a fixed $X \in \mathfrak{h}$, $M(\exp(s+t)X) \leq M(\exp(sX))M(\exp(tX))$. Hence, there are positive constants A, B such that for all $X \in \mathfrak{p}$, we have $M(\exp(X)) \leq Ae^{B\|X\|}$, and so $M(\exp(X)) \leq Ae^{Bd(\exp(X)[e],[e])}$. In other words, viewing $M(\cdot)$ as a function on H/K, we have for all $z \in H/K$ that

$$M(z) \leq A \exp(Bd(z,[e])).$$

However, it is well-known that $\varphi:L(N) \to \mathbb{R}$ given by $\varphi(X) = d(\exp(X)[e],[e])$ satisfies $\varphi(X) \leq \log|p(\|X\|)|$ for some polynomial p, and this implies the result.

We remark that in the above proof we have no a priori control over A, B. In particular, while p is determined just by the group H, the degree of the polynomial growth on f that we have obtained depends upon B. However, if Conjecture 12.2 was correct, this would imply that one has polynomial growth of degree determined only by the rank of E. For reference, we record this fact.

Proposition 12.5. Assume the hypotheses and conclusions of 12.2. Then $f: L(N) \to \mathbb{R}$ (as defined in Theorem 12.4) has polynomial growth of degree bounded by q, where $q = \text{rank}(E) - 1$.

We now return to actions of discrete groups. Let Γ, H be as in 3.1. Suppose Γ acts on an n-manifold M.

Definition 12.6. Let $L \subset H$ be a closed subgroup. The action of Γ on M is called L-distal if $p^*(TM) \to M \times H/L$ admits a Γ-invariant reduction to a distal subgroup of $GL(n,\mathbb{R})$, where $p: M \times H/L \to M$ is projection.

Of course if $L = H$, this is nothing more than the assertion that there is a Γ-invariant distal G-structure on M, and this clearly implies L-distality for any L. On the other hand, it is clear that for $L = \{e\}$, any action is L-distal.

Conjecture 12.7. Let Γ, H be as in 3.1 $(S = \{\infty\})$. Suppose Γ acts smoothly on a compact manifold M, preserving a smooth volume density. Then the action is N-distal where N is the unipotent radical of the minimal parabolic subgroup of H.

This is obviously in the same spirit as Conjecture 12.2, and in fact they are closely related. We recall that for any Γ-space Y, $H/\Gamma \times_\Gamma Y$ is the H-space $(H \times Y)/\Gamma$, which we sometimes call the induced H-space. Now suppose $V \to M$ is any vector bundle on which Γ acts by vector bundle automorphisms. Then $H/\Gamma \times_\Gamma V \to H/\Gamma \times_\Gamma M$ is a vector bundle (with the same fiber as $V \to M$) on which the induced H-action is by vector bundle automorphisms. The existence of an N-invariant distal reduction of this bundle is clearly equivalent to the existence of an H-invariant distal reduction of $q^*(H/\Gamma \times_\Gamma V) \to (H/\Gamma \times_\Gamma M) \times H/N$ where $q: (H/\Gamma \times_\Gamma M) \times H/N \to H/\Gamma \times_\Gamma M$ is the projection. However, it is easy to see that we have an isomorphism of vector bundles

$$\begin{array}{ccc} H/\Gamma \times_\Gamma (p^*V) & \cong q^*(H/\Gamma \times_\Gamma V) \\ \downarrow & \downarrow \\ H/\Gamma \times_\Gamma (M \times H/N) & \cong (H/\Gamma \times_\Gamma M) \times H/N \end{array}$$

commuting with the H-actions. (Here p is as in Def. 12.6 with $L = N$.) However, the existence of an H-invariant distal reduction of $H/\Gamma \times_\Gamma p^*(V) \to H/\Gamma \times_\Gamma (M \times H/N)$ is equivalent to the existence of a Γ-invariant distal reduction of $p^*(V) \to M \times H/N$. Hence we have established:

Proposition 12.8. If $V \to M$ is a vector bundle on which Γ acts by vector bundle automorphisms, and $p: M \times H/\Gamma \to M$ is the projection, then there is a Γ-invariant distal structure on the vector bundle $p^*(V) \to M \times H/N$ if and only if the H action on $H/\Gamma \times_\Gamma V \to H/\Gamma \times_\Gamma M$ is N-distal.

In particular:

Corollary 12.9. For Γ cocompact, Conjecture 12.2 implies Conjecture 12.7. If Γ is not cocompact, the generalization of 12.2 to manifolds with finite H-invariant volume implies 12.7.

In Section 9, we established Conjecture I for actions preserving a distal structure. In the next two sections, we shall see that via estimates of Howe and Harish-Chandra on matrix coefficients of unitary representations, one can establish Conjecture I for N-distal actions at least for cocompact lattices and in a suitable dimension range.

13. Results of Howe and Harish-Chandra on asymptotics of matrix coefficients of unitary representations

In this section we recall some fundamental results of Howe [5] and Harish-Chandra [3] on matrix coefficients. Once again, for simplicity, we restrict our attention to semisimple Lie groups, rather than groups over general local fields.

Let H be a semisimple Lie group with finite center, and π a unitary representation of H on a Hilbert space $\mathcal{H}$. A matrix coefficient of π is a function $f:H \to \mathbb{C}$ of the form $f(h) = \langle\pi(h)v,w\rangle$ for some $v,w \in \mathcal{H}$. Fix a maximal compact subgroup $K \subset H$.

Definition 13.1 (Howe [5]). If $\Phi:H \to \mathbb{R}$, we say that π is Φ-bounded if for all K-invariant $v,w \in \mathcal{H}$, the matrix coefficient $f(h) = \langle\pi(h)v,w\rangle$ satisfies $|f(h)| \leq |\Phi(h)| \|v\| \|w\|$.

Thus, Φ-boundedness is a condition of uniform behavior on K-invariant vectors. Howe discusses the more general case of K-finite vectors, but we will not need this here.

We recall that Harish-Chandra has considered the matrix coefficient $\Xi(h) = \langle\sigma(h)1,1\rangle$ where $\sigma = \mathrm{ind}_P^H(I)$, $P \subset H$ is the minimal parabolic subgroup, and 1 is the unique K-invariant unit vector for σ. Let $\mathfrak{h}$ be the Lie algebra of H, $\mathfrak{a} \subset \mathfrak{h}$ a split Cartan subalgebra, and $\mathfrak{a}^+$ the positive Weyl chamber with respect to a set of positive roots Σ^+. Let Δ_p be the modular function of P, (so $\Delta_p:P \to \mathbb{R}^+$). Then for $X \in \mathfrak{a}$,

$$\Delta_p(\exp X) = \exp\Big(\sum_{\alpha\in\Sigma^+} m(\alpha)|\alpha(X)|\Big) ,$$

where $m(\alpha) \in \mathbb{Z}^+$ is the multiplicity of α. We then have the following basic estimate.

Theorem 13.2 (Harish-Chandra [3]). Given $\varepsilon > 0$, there is $D > 0$ such that for all $X \in \mathfrak{a}^+$,

$$\Xi(\exp X) \leq D\Delta_p^{-1/(2+\varepsilon)}(\exp X).$$

A unitary representation π is called L^p if there is a dense subspace $\mathcal{H}_0 \subset \mathcal{H}$ such that the matrix coefficient defined by any $v,w \in \mathcal{H}_0$ is in $L^p(H)$. We then have:

Theorem 13.3 (Howe, [5, Corollary 7.2]). Suppose π is L^p where $p \leq 2m$ for some $m \in \mathbb{Z}^+$. Then π is $\Xi^{1/m}$-bounded.

It is easy to see that if $\pi = \int^{\oplus} \pi_t$ and each π_t is Φ-bounded, so is π. We thus have

Corollary 13.4. Suppose for every $\sigma \in \hat{H}$ (the unitary dual of H),

$\sigma \neq I$, that σ is L^p for some fixed p. Assume $p \leq 2m$ for some $m \in \mathbb{Z}^+$. Then any unitary representation π of H with no invariant vectors is $\Xi^{1/m}$-bounded.

The utility of this is indicated by another result of Howe.

Theorem 13.5 (Howe). Assume $\mathbb{R}$-rank$(H') \geq 2$ for every simple factor of H. Then there is some p such that every non-trivial $\sigma \in \hat{H}$ is L^p.

In our applications in the next section to actions of discrete groups we shall need quantitative information on p.

Theorem 13.6 (Howe [5, Theorem 8.4] ff.). Let $Sp(2m,\mathbb{R})$ be the symplectic group (for a symplectic form on a 2m-dimensional space.) Assume $m > 2$. Then for all $\varepsilon > 0$, every non-trivial $\sigma \in Sp(2m,\mathbb{R})^\wedge$ is $L^{2m+\varepsilon}$.

Corollary 13.7. If $m > 2$, every unitary representation of $Sp(2m,\mathbb{R})$ with no (non-trivial) invariant vectors is $\Xi^{1/(m+1)}$-bounded.

Proof. 13.6, 13.4.

Combining 13.7 and 13.2, we obtain the following result.

Proposition 13.8. Assume $m > 2$. Then there is a unipotent 1-parameter subgroup $u_t \in Sp(2m,\mathbb{R})$ such that for all $\varepsilon > 0$, there is a positive constant C such that for any unitary representation π of $Sp(2m,\mathbb{R})$ with no (non-trivial) invariant vectors, any matrix coefficient f defined by two K-invariant unit vectors $(K \subset Sp(2m,\mathbb{R})$ the maximal compact subgroup) satisfies

$$|f(u_t)| \leq C(t)^{-[(m^2-m)/(m+1)]+\varepsilon} \quad \text{for } t \text{ sufficiently large.}$$

Proof. Following Howe's notation in [5], we let $\{e_i,f_i\}$, $i = 1,\ldots,m$, be a basis of $\mathbb{R}^{2m}$ such that $B(e_i,e_j) = 0$, $B(f_i,f_j) = 0$ and $B(e_i,f_j) = \delta_{ij}$ where B is the symplectic form. Let X be the span of $\{e_i\}$, and Y the span of $\{f_i\}$. Identifying $GL(X)$ and $GL(Y)$ with $GL(m,\mathbb{R})$ via these bases, we have a map $GL(m,\mathbb{R}) \to Sp(2m,\mathbb{R})$

such that $A \in GL(m,\mathbb{R})$ acts by A on X and ${}^t(A^{-1})$ on Y. Let $x \in \mathfrak{gl}(X)$ be the linear map with $x(e_i) = e_{i-1}, x(e_1) = 0$. Then $u_t = \exp(tx)$ is a one parameter unipotent subgroup of $GL(X) \cong GL(n,\mathbb{R}) \hookrightarrow Sp(2m,\mathbb{R})$.

Let $a_t \in GL(X)$ be the diagonal term in the KAK-decomposition of u_t in $GL(X)$. Then we clearly have for some (positive definite) $O(n,\mathbb{R})$-invariant inner product on X that $\|(u_t^* u_t)^{1/2}\| \geq |p(t)|$ where p is a polynomial of degree m-1, and hence that $t > 0$ implies $\|a_t\| \geq |p(t)|$. In particular, the maximal absolute value of the eigenvalues of a_t is at least $|p(t)|$. It follows from the formula for Δ_p (see the paragraph preceding Theorem 13.2, and for $Sp(2m,R)$ specifically, see [5, p. 313]) that viewing $u_t, a_t \in Sp(2m,\mathbb{R})$, there is an element w in the Weyl group such that

$$\Delta_p(wa_tw^{-1}) \geq |p(t)|^{2m}.$$

By Theorem 13.2 and the K-bi-invariance of Ξ, we deduce that for all $\varepsilon > 0$, there is $C > 0$ such that $\Xi(u_t) \leq C|p(t)|^{2m((-1/2)+\varepsilon)}$, and hence (changing C if necessary), that $\Xi(u_t) \leq C|p(t)|^{-m+\varepsilon}$. Since p is a polynomial of degree m-1, the result follows from Corollary 13.7.

14. Applications to cocompact lattices acting on low-dimensional manifolds.

The point of this section is to prove the following theorem.

Theorem 14.1. For each integer $n \in \mathbb{Z}$, $n \geq 2$, there is an integer $\ell(n)$ with the following property. Suppose H is a real algebraic simple Lie group with $\mathbb{R}$-rank$(H) \geq \ell(n)$. Then for any compact manifold M, with $\dim M \leq n$ and any cocompact lattice $\Gamma \subset H$, there is no volume preserving ergodic action of Γ on M which is N-distal. (Def. 12.6.)

Remarks. 1) We recall the conjecture (12.7) that every volume preserving action of Γ is N-distal. Thus, 14.1 shows that a proof of Conjecture 12.7 would yield great progress on the proof of Conjecture I.

2) The function $\ell(n)$ is effectively computable from the proof. We discuss this following the proof.

To prove Theorem 14.1, we shall make use of Proposition 13.8 via symplectic subgroups of H. A case by case analysis easily shows the following.

Lemma 14.2. There is a function $q:\mathbb{Z}^+ \to \mathbb{Z}^+$ such that any real algebraic simple Lie group H with $\mathbb{R}$-rank $H \geq q(m)$ admits a rational homomorphism $\pi: Sp(2m,\mathbb{R}) \to H$ which is a local isomorphism onto its image.

The function q is of course effectively computable. Theorem 14.1 then follows from Lemma 14.2 and the following theorem.

Theorem 14.3. There is a function $\mu:\mathbb{Z}^+ \to \mathbb{Z}^+$ with the following property. If H is a real algebraic simple Lie group admitting a rational homomorphism $Sp(2m,\mathbb{R}) \to H$ which is a local isomorphism onto its image where $m \geq \mu(n)$, then for compact manifold M with $\dim M \leq n$, and any cocompact lattice $\Gamma \subset H$, there is no volume preserving ergodic action of Γ on M which is N- distal. The function μ is effectively (and easily) computable.

The function ℓ in Theorem 14.1 is just $q \circ \mu$. Theorem 14.3 is of course more precise than 14.1, and for a given H, 14.3 can be applied directly rather than using 14.1.

Proof of Theorem 14.3. Fix $r \geq 1$ and consider the bundle maps $J^r(P(M);\mathbb{R}) \to P(M) \to M$, the first being a vector bundle, the second a principal bundle. Let $V = H/\Gamma \times_\Gamma J^r(P(M);\mathbb{R})$, $Q = H/\Gamma \times_\Gamma P(M)$, and $X = H/\Gamma \times_\Gamma M$ (with notation as in Section 12.) The Γ-maps above then induce natural H-maps $V \overset{\pi}{\to} Q \overset{P}{\to} X$. Here π is a vector bundle projection with fiber $J^r(\mathbb{R}^\ell,0;\mathbb{R})$ where $\ell = \dim P(M)$ and $Q \to X$ is a principal $GL(n,\mathbb{R})$-bundle. The following lemma is not hard to verify and we omit the proof.

Lemma 14.4. If the Γ-action on M is N-distal, then there is a distal subgroup $G \subset GL(n,\mathbb{R})$ and i) an N-invariant reduction of $Q \to X$ to a principal G-bundle $Q_0 \to X$; and ii) an $N \times G$-invariant distal structure on the vector bundle $V \to Q$.

By Corollaries 4.6, 4.13 for $n < d(H)$ (and in particular for $n < d(Sp(2m,\mathbb{R}))$, there is a measurable $\Gamma \times GL(n,\mathbb{R})$-invariant metric

η on $J^r(P(M);\mathbb{R}) \to P(M)$. This induces a measurable $H \times GL(n,\mathbb{R})$-invariant metric $\tilde{\eta}$ on $V \to Q$.

Lemma 14.5. To prove Theorem 14.3, it suffices to prove that there is a function μ' so that for $m \geq \mu'(n)$, $\dim M \leq n$, we have for $r = \dim P(M) + 1$ that $M(\tilde{\eta}/\xi)$, $M(\xi/\tilde{\eta}) \in L^2_{loc}(Q)$ for any one (and hence all) smooth metrics ξ on $V \to Q$.

Proof. From Fubini's theorem, Theorem 7.1, Theorem 6.1 (and the remarks following its proof), we deduce the existence of a function μ_0 such that for $m \geq \mu_0(n)$, $\dim M \leq n$, we have that any such Γ-action preserves a C^0-metric. If we let I be the compact group of isometries of M with respect to the topological distance function defined by this C^0-metric, we have that I acts continuously and transitively on M, since Γ acts ergodically (and hence with at least one dense orbit) on M. Therefore I is a Lie group [10, p. 244], and $\dim I \leq n(n+1)/2$. Write $I = AI'$ where A, I' are normal subgroups, A is abelian, and I' is semisimple. Since Γ is Kazhdan, any homomorphism of Γ into I/I' has finite image, and the lemma then follows from Theorem 3.8.

We now choose the smooth metric ξ on $V \to Q$ with a good invariance property.

Lemma 14.6. Let $K \subset Sp(2m,\mathbb{R})$ be a maximal compact subgroup. Then there is a $K \times GL(n,\mathbb{R})$-invariant smooth metric ξ on the vector bundle $V \to Q$. (Here $Sp(2m,\mathbb{R})$ acts via the local embedding $Sp(2m,\mathbb{R}) \to H$).

Proof. Since K is compact and the action commutes with that of $GL(n,\mathbb{R})$, by the standard averaging argument it suffices to construct a smooth $GL(n,\mathbb{R})$-invariant metric. Choose any reduction of $Q \to X$ to $O(n,\mathbb{R})$, say $Q_1 \subset Q$. There is clearly an $O(n,\mathbb{R})$-invariant smooth metric on V over the subbundle $Q_1 \subset Q$. This can clearly be extended to a smooth $GL(n,\mathbb{R})$-invariant metric on V defined over all Q. (Cf. the discussion preceding Proposition 2.4).

Now let ξ be a smooth $K \times GL(n,\mathbb{R})$-invariant metric on $V \to Q$, and $\tilde{\eta}$ the measurable $H \times GL(n,\mathbb{R})$-invariant metric defined above. The functions $M(\tilde{\eta}/\xi)$, $M(\xi/\tilde{\eta})$ on Q are clearly $GL(n,R)$-invariant and hence factor to measurable functions $\bar{M}(\tilde{\eta}/\xi)$, $\bar{M}(\xi/\tilde{\eta})$ on X. From Lemma 14.5 we have:

Lemma 14.7. It suffices to prove, in the context of Lemma 14.5, that $\bar{M}(\tilde{\eta}/\xi)$, $\bar{M}(\xi/\tilde{\eta}) \in L^2(X)$.

Lemma 14.8. Let d be the rank of the vector bundle $V \to Q$ (i.e. $d = \dim J^{\dim P+1}(\mathbb{R}^{\dim P},0;\mathbb{R})$), and $u \in N$. Then there is a polynomial p with $\deg(p) = d-1$ and such that for all $x \in X$ and $n \in \mathbb{Z}^+$,

$$\bar{M}(\tilde{\eta}/\xi)(u^n x) \leq |p(n)|\bar{M}(\tilde{\eta}/\xi)(x).$$

The same assertion is true for the function $\bar{M}(\xi/\tilde{\eta})$.

Proof. The proof is similar in spirit to the proof of Theorem 9.2. We fix a compact set $Q_1 \subset Q_0$ (where Q_0 is as in Lemma 14.4) such that Q_1 projects surjectively onto X. Let $\tilde{Q}_1 = Q_1 \cup uQ_1$, so that $\tilde{Q}_1$ is also compact. We can choose a measurable framing of the bundle $V \to Q$ on the set $\tilde{Q}_1$ such that:

i) at each point in $\tilde{Q}_1$, the frame is contained in the $N \times G$-invariant distal structure on $V \to Q$ given by Lemma 14.4 (ii); (let $L \subset GL(d,\mathbb{R})$ be the corresponding distal subgroup);

and, ii) If $\tilde{\xi}$ is the measurable metric on $V \to Q$ defined over the set $\tilde{Q}_1$ given by this measurable framing, then $M(\tilde{\xi}/\xi)$, $M(\xi/\tilde{\xi})$ are uniformly bounded in $\tilde{Q}_1$.

As in the calculation in the proof of 9.2, we see that for any $y \in Q$, $n \in \mathbb{Z}^+$, and $g \in G$, we have

$$M(\tilde{\eta}/\xi)(u^n yg) = M(\tilde{\eta}/\xi)(y)M(\xi/d(u^n,g)^*\xi)(y).$$

Each metric ξ_y, $y \in \tilde{Q}_1$ can be expressed as a matrix in terms of the measurable framing described above. Since $M(\tilde{\xi}/\xi)$, $M(\xi/\tilde{\xi})$ are uniformly bounded on $\tilde{Q}_1$, $\{\xi_y | y \in \tilde{Q}_1\}$ is a precompact set of matrices. It therefore suffices to see that there is a polynomial p with $\deg(p) = d-1$ such that for each $y \in Q_1$, there exists $g \in G$ such that $u^n yg \in Q_1$ and $\|d(u^n,g)_y^{-1}\| \leq p(n)$ where the norm is taken with respect to this measurable framing.

The argument now proceeds as in the proof of Theorem 9.2. We deduce that there is a compact symmetric neighborhood of the identity $W \subset L$ such that (with respect to the measurable framing), for each

$y \in Q_1$ and $n \in \mathbb{Z}^+$, there is some $g \in G$ with $u^n yg \in Q_1$ and $d(u^n,g)_y \in W^n$. Since $L \subset GL(d,\mathbb{R})$ is a distal linear group, $\max\{\|\ell\| \mid \ell \in W^n\} \leq |p(n)|$ where $\deg(p) = d-1$. This proves the lemma.

Proof of Theorem 14.3. Let $\varphi = \bar{M}(\tilde{\eta}/\xi)$, $\tilde{\varphi} = \bar{M}(\xi/\tilde{\eta})$, so $\varphi,\tilde{\varphi}:X \to \mathbb{R}$. By Lemma 14.7, it suffices to show a function μ' exists so that with the corresponding hypotheses, $\varphi,\tilde{\varphi} \in L^2(X)$. We give the proof for φ as a similar argument applies to $\tilde{\varphi}$. We have the following properties of φ:

i) φ is K-invariant (since ξ and $\tilde{\eta}$ are; see Lemma 14.6).

ii) For any $u \in N$, $n \in \mathbb{Z}^+$, and $x \in X$, we have $|\varphi(u^n x)| \leq |p(n)||\varphi(x)|$ where p is a polynomial of degree $d-1$. (Here d is as in 14.8 and hence depends only on $\dim M$.)

We now claim that any φ satisfying (i), (ii) is in $L^2(X)$ as long as m is sufficiently large, depending only on $\dim M$. Namely, choose m such that $(m^2-m)/(m+1) > d-1$. We identify the element $u_1 \in Sp(2m,\mathbb{R})$ in the 1-parameter subgroup u_t of Proposition 13.8 with an element $u \in H$. Then by condition (ii) above and Proposition 13.8 we can choose n sufficiently large such that setting $h = u^n$ (which we view as an element of $Sp(2m,\mathbb{R})$ and H) we have

a) For some $B > 0$, we have $|\varphi(hx)| \leq B|\varphi(x)|$ for all $x \in X$;

b) for any unitary representation π of $Sp(2m,\mathbb{R})$ with no invariant vectors, any matrix coefficient f defined by two K-invariant vectors unit vectors satisfies $|f(h)| \leq \varepsilon$, where $\varepsilon B^2 < 1$.

We will now apply the same type of integrability argument we used in Section 8. The following is simply a reformulation of Lemma 8.3.

Lemma 14.9. Let (X,μ) be a probability space, and $A \subset X$ measurable with $\mu(A)(1-\mu(A)) > 0$. Let f_A be the normalization of the projection of χ_A onto $L^2(X) \ominus \mathbb{C}$. If $\mu(A) = \mu(B)$ and $|\langle f_A,f_B\rangle| < \varepsilon$, then

$$\mu(A \cap (X-B)) \geq (1-\varepsilon)\mu(A)(1-\mu(A)).$$

Continuing with the proof of Theorem 14.3, choose $0 < w < 1$ such that $[1-(1-\varepsilon)w]B^2 < 1$. (This is possible since $\varepsilon B^2 < 1$.) Let $A_0 \subset X$ be a measurable set such that:

i) A_0 is K-invariant.

ii) $\varphi|A_0$ is bounded, say $|\varphi(x)| \leq C$ for $x \in A_0$.

iii) $\mu(A_0) = w$.

(We remark that every K-orbit in X is a null set, and hence inside any K-invariant measurable set there is a K-invariant measurable set with arbitrary smaller measure.) We can obtain (i) since φ is K-invariant. Since Γ acts ergodically on M, H acts ergodically on X (preserving a finite volume), and therefore $Sp(2m,\mathbb{R})$ also acts ergodically by Moore's theorem [30] (or by the vanishing at infinite of matrix coefficients for simple groups [30].) Let π be the associated representation of $Sp(2m,\mathbb{R})$ on $L^2(X) \ominus \mathbb{C}$. By property (b) above, $|\langle\pi(h)f_{A_0}, f_{A_0}\rangle| < \varepsilon$. Lemma 14.9 then implies that

$\mu(hA_0 \cap (X-A_0)) \geq (1-\varepsilon)w(1-\mu(A_0))$. Let $A_1 \subset K(hA_0 \cap (X-A_0))$ be a K-invariant set with $\mu(A_1) = (1-\varepsilon)w(1-\mu(A_0))$. As in Section 8, we repeat this procedure applied to the K-invariant set $A_0 \cup A_1$. Continuing inductively, we find a sequence of mutually disjoint measurable K-invariant sets A_i, $i \geq 0$, such that $X = \cup_{i=0}^{\infty} A_i$, for $x \in A_n$ we have $|\varphi(x)| \leq B^n C$, and setting $a_n = \mu(A_n)$, we have (as in Section 8) that

$$a_{n+1} = (1-\varepsilon)w(1 - \sum_{i=0}^{n} a_i).$$

Thus, for $n \geq 0$, $a_{n+1} = a_n[1 - (1-\varepsilon)w]$. We have

$$\int|\varphi|^2 \leq \sum_{n=0}^{\infty} B^{2n}C^2 a_n.$$

This sum is finite since $B^2[1 - (1-\varepsilon)w] < 1$. This completes the proof of Theorem 14.3.

Remark. An examination of the proof shows that for a given n it suffices to choose $m = \mu(n)$ so that the following (redundant) conditions hold:

i) $n < 2m$ (to apply superrigidity for cocycles in the form of Corollary 4.6).

ii) $\frac{n(n+1)}{2} < \dim Sp(2m,\mathbb{R})$ (to ensure the non-existence of ergodic actions preserving a C^0-metric using Theorem 3.8.)

iii) $(d-1) < (m^2-m)/(m+1)$ (to obtain the comparison of growth estimates (a), (b) in the proof above), where $d = \dim J^{r+1}(\mathbb{R}^r, 0; \mathbb{R})$ and $r = n^2 + n = \dim P(M)$.

The estimate in (iii) can be improved in a very simple way to $r = n^2 - 1+n$ by working with the bundle of special frames since we are assuming the existence of an invariant volume density. However, it can probably be improved in a more serious manner, since in the assertion of Lemma 14.8 the degree of the polynomial p was based on the growth rate for the maximal unipotent group in $GL(d,\mathbb{R})$, while the relevant unipotent group is really a subgroup of a prolongation of a smaller general linear group, the prolongation being naturally embedded in $GL(d,\mathbb{R})$. Another improvement should be possible in the rational function of m that appears in (iii). Namely, a more careful examination of the construction in the proof of Proposition 13.8 should give a stronger estimate than is presented here. In this way one can hopefully obtain some better control over the rather drastic growth of $\mu(n)$ required by (iii). Both of the suggestions above should be fairly routine to carry out.

15. A topological epilogue

For each n, we can consider the standard embedding $SL(n,\mathbb{Z}) \hookrightarrow SL(n+1,\mathbb{Z})$. We then set $SL(\mathbb{Z}) = \cup_{n=2}^{\infty} SL(n,\mathbb{Z})$. A natural conjecture in light of the results of this paper is that any action of $SL(\mathbb{Z})$ on a compact manifold is in some sense trivial. The following unpublished result of S. Weinberger verifies this.

Theorem 15.1 (S. Weinberger). Every action of $SL(\mathbb{Z})$ by homeomorphisms of a compact manifold is trivial.

The proof uses the existence of large finite subgroups of $SL(\mathbb{Z})$ and techniques from the topological theory of finite transformation groups.

REFERENCES

1. H. Abels, Which groups act distally?, Eng. Th. and Dyn. Sys.

2. V. Guillemin, S. Sternberg, Deformation theory of pseudogroup structures, Mem. Amer. Math. Soc., no. 64, 1966.

3. Harish-Chandra, Discrete series for semisimple Lie groups, II, Acta. Math., 116 (1966), 1-111.

4. S. Helgason, Differential Geometry and Symmetric Spaces. Academic Press, New York, 1962.

5. R. Howe, On a notion of rank for unitary representations of the classical groups, in Harmonic Analysis and Group Representations, ed. A. Figa-Talamanca, Liguori, Naples, 1982.

6. D. Kazhdan, Connection of the dual space of a group with the structure of its closed subgroups, Funct. Anal. Appl., 1

(1967), 63-65.

7. S. Kobayashi, Transformation groups in differential geometry, Springer, New York, 1972.

8. J.L. Koszul, Lectures on Transformation Groups, Tata Institute Lectures on Mathematics and Physics, no. 20, 1965.

9. G.A. Margulis, Discrete groups of motions of manifolds of non-positive curvature, A.M.S. Translations, 109 (1977), 33-45.

10. D. Montgomery, L. Zippin, Topological Transformation Groups, Interscience, New York, 1955.

11. C.C. Moore, Distal affine transformation groups, Amer. J. Math., 90 (1968), 733-751.

12. C.C. Moore, R.J. Zimmer, Groups admitting ergodic actions with generalized discrete spectrum, Invent. Math., 51 (1979), 171-188.

13. G.D. Mostow, Strong rigidity of locally symmetric spaces, Annals of Math. Studies, no. 78 (1973).

14. R. Palais, Seminar on the Atiyah-Singer index theorem, Annals of Math. Studies, no. 57.

15. R. Palais, Foundations of Global Non-Linear Analysis, Benjamin, New York, 1968.

16. B. Reinhart, Differential Geometry of Foliations, Springer, New York.

17. G. Reeb, P.A. Schweitzer, W. Schachermayer, Lecture Notes in Math. no. 652, 138-140.

18. S. Sternberg, Lectures on Differential Geometry, Prentice-Hall, Englewood Cliffs, N.J., 1964.

19. C.L. Terng, Natural vector bundles and natural differential operators, Amer. J. Math., 100 (1978), 775-828.

20. W. Thurston, A generalization of the Reeb stability theorem, Topology, 13 (1974).

21. S. Weinberger, unpublished.

22. R.J. Zimmer, Extensions of ergodic group actions, Ill. J. Math., 20 (1976), 373-409.

23. R.J. Zimmer, Strong rigidity for ergodic actions of semisimple Lie groups, Annals of Math., 112 (1980), 511-529.

24. R.J. Zimmer, Volume preserving actions of lattices in semisimple groups on compact manifolds, Publ. Math. I.H.E.S., 59 (1984), 5-33.

25. R.J. Zimmer, Actions of lattices in semisimple groups preserving a G-structure of finite type, Erg. Th. and Dyn. Sys., to appear.

26. R.J. Zimmer, Lattices in semisimple groups and distal geometric structures, Invent. Math., 80 (1985), 123-137.

27. R.J. Zimmer, Kazhdan groups acting on compact manifolds, Invent. Math., 75 (1984), 425-436.

28. R.J. Zimmer, Semisimple automorphism groups of G-structures, J. Diff. Geom., 19 (1984), 117-123.

29. R.J. Zimmer, On the automorphism group of a compact Lorenz manifold and other geometric manifolds, Invent. Math. 83 (1986), 411-424.

30. R.J. Zimmer, Ergodic Theory and Semisimple Groups, Birkhauser-Boston, Cambridge, 1984.

31. R.J. Zimmer, Ergodic Theory and the automorphism group of a G-structure, preprint.

32. R.J. Zimmer, On discrete subgroups of Lie groups and elliptic geometric structures, Rev. Ibero-Amer., vol. 1, 1985.

DEPARTMENT OF MATHEMATICS
UNIVERSITY OF CHICAGO
CHICAGO, IL 60637

Progress in Mathematics

1 Gross. Quadratic Forms in Infinite-Dimensional Vector Spaces
2 Pham. Singularités des Systèmes Différentiels de Gauss-Manin
3 Okonek/Schneider/Spindler. Vector Bundles on Complex Projective Spaces
4 Aupetit. Complex Approximation, Proceedings, Quebec, Canada, July 3-8, 1978
5 Helgason. The Radon Transform
6 Lion/Vergne. The Weil Representation, Maslov Index and Theta Series
7 Hirschowitz. Vector Bundles and Differential Equations Proceedings, Nice, France, June 12-17, 1979
8 Guckenheimer/Moser/Newhouse. Dynamical Systems, C.I.M.E. Lectures, Bressanone, Italy, June, 1978
9 Springer. Linear Algebraic Groups
10 Katok. Ergodic Theory and Dynamical Systems I
11 Balslev. 18th Scandinavian Congress of Mathematicians, Aarhus, Denmark, 1980
12 Bertin. Séminaire de Théorie des Nombres, Paris 1979-80
13 Helgason. Topics in Harmonic Analysis on Homogeneous Spaces
14 Hano/Marimoto/Murakami/Okamoto/Ozeki. Manifolds and Lie Groups: Papers in Honor of Yozo Matsushima
15 Vogan. Representations of Real Reductive Lie Groups
16 Griffiths/Morgan. Rational Homotopy Theory and Differential Forms
17 Vovsi. Triangular Products of Group Representations and Their Applications
18 Fresnel/van der Put. Géometrie Analytique Rigide et Applications
19 Oda. Periods of Hilbert Modular Surfaces
20 Stevens. Arithmetic on Modular Curves
21 Katok. Ergodic Theory and Dynamical Systems II
22 Bertin. Séminaire de Théorie des Nombres, Paris 1980-81
23 Weil. Adeles and Algebraic Groups
24 Le Barz/Hervier. Enumerative Geometry and Classical Algebraic Geometry
25 Griffiths. Exterior Differential Systems and the Calculus of Variations
26 Koblitz. Number Theory Related to Fermat's Last Theorem
27 Brockett/Millman/Sussman. Differential Geometric Control Theory
28 Mumford. Tata Lectures on Theta I
29 Friedman/Morrison. Birational Geometry of Degenerations
30 Yano/Kon. CR Submanifolds of Kaehlerian and Sasakian Manifolds
31 Bertrand/Waldschmidt. Approximations Diophantiennes et Nombres Transcendants
32 Books/Gray/Reinhart. Differential Geometry
33 Zuily. Uniqueness and Non-Uniqueness in the Cauchy Problem
34 Kashiwara. Systems of Microdifferential Equations
35 Artin/Tate. Arithmetic and Geometry: Papers Dedicated to I.R. Shafarevich on the Occasion of His Sixtieth Birthday, Vol. I
36 Artin/Tate. Arithmetic and Geometry: Papers Dedicated to I.R. Shafarevich on the Occasion of His Sixtieth Birthday, Vol. II
37 de Monvel. Mathématique et Physique
38 Bertin. Séminaire de Théorie des Nombres, Paris 1981-82
39 Ueno. Classification of Algebraic and Analytic Manifolds
40 Trombi. Representation Theory of Reductive Groups